高校教师人工智能素养红皮书

（2024 年版）

Red Book on Artificial Intelligence Literacy of University Educators

（2024 Edition）

浙江大学人工智能教育教学研究中心　编著

Zhejiang University Research Center of
Artificial Intelligence for Education and Teaching

ZHEJIANG UNIVERSITY PRESS
浙江大学出版社
·杭州·

图书在版编目（CIP）数据

高校教师人工智能素养红皮书：2024年版 / 浙江大学人工智能教育教学研究中心编著. --杭州：浙江大学出版社，2025. 6. -- ISBN 978-7-308-26360-3

Ⅰ．TP18

中国国家版本馆 CIP 数据核字第 2025D3C495 号

高校教师人工智能素养红皮书：2024年版
GAOXIAO JIAOSHI RENGONG ZHINENG SUYANG HONGPISHU: 2024 NIAN BAN

浙江大学人工智能教育教学研究中心　编著

策划编辑	黄娟琴　李　晨
责任编辑	陈丽勋　柯华杰
责任校对	郑成业
封面设计	林智广告
出版发行	浙江大学出版社
	（杭州市天目山路148号　邮政编码 310007）
	（网址：http://www.zjupress.com）
排　　版	杭州林智广告有限公司
印　　刷	杭州捷派印务有限公司
开　　本	710mm×1000mm　1/16
印　　张	9
字　　数	120千
版 印 次	2025年6月第1版　2025年6月第1次印刷
书　　号	ISBN 978-7-308-26360-3
定　　价	40.00元

前　言

　　党的二十大报告强调"教育、科技、人才是全面建设社会主义现代化国家的基础性、战略性支撑"，"要坚持教育优先发展、科技自立自强、人才引领驱动，加快建设教育强国、科技强国、人才强国"。[①] 在 2024 年的全国教育大会上，习近平总书记再次强调教育是强国建设、民族复兴之基。建成教育强国是近代以来中华民族梦寐以求的美好愿望，是实现以中国式现代化全面推进强国建设、民族复兴伟业的先导任务、坚实基础、战略支撑，必须朝着既定目标扎实迈进。

　　近些年来，人工智能技术的迅猛发展及其日益广泛的行业应用正在引发人类社会的深刻变革，"人工智能＋"成为各行各业发展的重要趋势。推动人工智能研发及应用的核心在人才、根本在教育、关键在教师。生成式人工智能的出现更是深刻改变了人

①　习近平：《高举中国特色社会主义伟大旗帜 为全面建设社会主义现代化国家而团结奋斗——在中国共产党第二十次全国代表大会上的报告》，人民出版社2022年版，第33页。

类与环境互动的能力和角色，正彻底改变着以知识积累和传递为中心的教学模式，推动课堂教学模式从传统的"师—生"二元结构转向"师—机—生"三元结构，重塑教师在教育教学中的角色与定位，人机协同新形态正带动教育手段和教育机制的历史性变革。因此，人工智能教育应用对于高校而言不是一般的策略性问题，而是影响甚至决定教育高质量发展的战略性、全局性问题。

高校教师是智能技术能否，以及如何进入高校教学科研场景的关键守门人和把关者，也是高校智能教育能否开展，以及如何开展的设计者和实施者。因此，高校教师人工智能的素养水平将直接决定智能时代高校教育教学变革的方向、速度和质量。

本红皮书旨在提出高校教师人工智能素养的概念与内涵，以及提升的目标、路径与保障，认为高校教师人工智能素养是指在高校从事教学与科研工作的教师为了在智能时代胜任教书育人、科研创新、社会服务和文化传承等工作而应具有的与人工智能应用相关的专门素养。它包含赓续育人理念（何为师）、掌握智能知识（以何为师）、变革教研模式（何以成师）和担当社会责任（师者何为）等能力。具体而言，高校教师人工智能素养包括智能时代育人理念、智能教育基本知识、人机协同教学能力、数智赋能科研创新和科技向善人本价值等五个维度的内容。其中，理念引领、知识为基、能力为核、创新为重、价值为本，五者相辅相成、相互融合。

PREFACE

The report to the 20th National Congress of the Communist Party of China (CPC) emphasizes that "Education, science and technology, and human resources are the foundational and strategic pillars for building a modern socialist country in all respects" and that "We will continue to give high priority to the development of education, build China's self-reliance and strength in science and technology, and rely on talent to pioneer and to propel development. We will speed up work to build a strong educational system, greater scientific and technological strength, and a quality workforce." [1] At the 2024 National Conference on Education, General Secretary Xi Jinping once again emphasized the foundational role that education plays in building a great country and realizing national rejuvenation. He pointed out that building a leading country in education has been a long-cherished aspiration of the Chinese nation since the advent of modern times. It is a guiding task,

---

[1] Xi Jinping, "Hold High the Great Banner of Socialism with Chinese Characteristics and Strive in Unity to Build a Modern Socialist Country in All Respects—Report to the 20th National Congress of the Communist Party of China," Xinhua News Agency, 2022, http://english.www.gov.cn/news/topnews/202210/25/content_WS6357df20c6d0a757729e1bfc.html.

solid foundation and strategic support for building China into a great country and realizing national rejuvenation on all fronts through Chinese modernization. We must make solid progress toward this set goal.

Rapid development of artificial intelligence (AI) technology and its diverse applications across industries are driving significant transformations in society. AI+ has become an important trend in the development of various industries. The core of promoting the research and application of AI lies in talent, education, and educators. The emergence of Generative AI (GenAI) has profoundly changed the ability and role of human interaction with the environment, completely changing the teaching mode centered on knowledge accumulation and transmission, promoting the shift of classroom teaching mode from the traditional "educator–learner" binary structure to the "educator–AI–learner" ternary structure, reshaping the role and positioning of educators in education, and the new form of human–AI collaboration is driving a historic change in educational methods and mechanisms. Therefore, the application of AI in education is not a general strategic issue for higher education institutions (HEIs), but a strategic and global issue that affects and even determines the high-quality development of education.

University educators serve as the key gatekeepers and supervisors in determining how and to what extent intelligent technology could be integrated into teaching and research in HEIs. They are also the designers and practitioners of whether and how intelligent education can be carried out in HEIs. Therefore, the level of AI literacy of university educators will directly determine the direction, speed, and quality of the transformation of university education in the age of intelligence.

This red book aims to propose the concept and connotation of AI literacy of university educators, as well as the goals, pathways, and supports for its improvement. The AI literacy of university educators refers to the competencies they should possess in the age of intelligence. These include a commitment to the advanced concepts of education (what it means to be an educator), mastery of AI knowledge (what contributes to an educator), transformation of teaching and research models (how to evolve as an educator), and the ability to assume social responsibility (what it means to be a responsible educator). Specifically, AI literacy of university educators encompasses five key dimensions: advanced concepts in the age of intelligence, basic knowledge of intelligent education, human–AI collaborative teaching capability, scientific research and innovation empowered by AI, and human-centered values of technology for greater good. Among these, advanced concepts serve as the guide, knowledge as the foundation, capability as the core, innovation as the focus, and values as the essence.

目　录

第一部分

智能时代高校教师面临的变革

一、人工智能带给高校教师的挑战和机遇 ················· 003

二、智能时代高校教师的定位及技能要求 ················· 009

第二部分

高校教师人工智能素养的概念与内涵

一、高校教师人工智能素养的概念界定 ················· 015

二、高校教师人工智能素养的内涵解读 ················· 019

第三部分

高校教师人工智能素养提升的目标、路径与保障

一、高校教师人工智能素养提升的目标 ················· 031

二、高校教师人工智能素养提升的路径 ················· 031

三、高校教师人工智能素养提升的保障 ················· 037

结　语 ················· 040

附　录

附录 1：与人工智能有关的高校虚拟教研室名单及案例·········042

附录 2：国内外代表性高校生成式人工智能使用指南名录······046

附录 3：《高等教育中的 ChatGPT 和人工智能：快速入门

指南》中的使用建议 ·······································053

CONTENTS

Part I

Transformations Faced by University Educators in the Age of Intelligence

Challenges and Opportunities of AI for Higher Education Educators...059

Positioning and Skill Requirements for University Educators in the Age of Intelligence 068

Part II

The Concept and Connotation of AI Literacy of University Educators

Definition of AI Literacy of University Educators 075

Connotation of AI Literacy of University Educators.......082

Part III

Goals, Pathways, and Supports for Enhancing University Educators' AI Literacy

Goals for Enhancing University Educators' AI Literacy........099

Pathways for Enhancing University Educators' AI
Literacy 099
Supports for Enhancing University Educators' AI
Literacy 108

Conclusion 113

Appendices
Appendix 1   List of AI-Related Virtual Teaching and
Research Studios in Higher Education and Case Study
.... 115
Appendix 2   List of Guidelines for the Use of GenAI in
Representative HEIs Worldwide 121
Appendix 3   Usage Suggestions in "ChatGPT and
Artificial Intelligence in Higher Education: Quick Start
Guide" 129

1

第一部分

智能时代高校教师
面临的变革

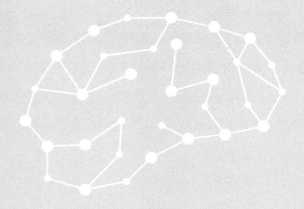

一、人工智能带给高校教师的挑战和机遇

人工智能正在成为一种通用目的技术，以史无前例的速度全方位改变着全球高等教育的发展方向和高校教师的教学及科研环境。

一方面，行业人才需求正在发生巨变。根据世界经济论坛发布的《未来就业报告2023》，至2027年，全球预计将失去8300万个传统工作岗位，同时将创造6900万个新的工作岗位。人工智能、机器学习专家等领域的岗位成为2023—2027年增长最为迅猛的工作岗位，人工智能与大数据能力被认定为未来最需要培养的技能之一。[①]2019年以来，我国人力资源和社会保障部同有关部门先后发布了五批共74个新职业，经梳理发现，这些新职业密集分布于智能信息技术相关行业中。由此可见，我国未来就业市场对于智能技术行业应用相关的人才需求也将持续增加。

鉴于人工智能行业应用的普遍性，联合国教科文组织（UNESCO）于2024年发布了在校师生人工智能能力框架，将

---

① World Economic Forum, "The Future of Jobs Report 2023," 2023, https://www.weforum.org/publications/the-future-of-jobs-report-2023/.

人工智能素养列为学生及教师必备素养，指出在校师生需要具备人工智能相关的知识、技能和态度，能够在教育及其他领域通过安全且有意义的方式理解和使用人工智能。[①]为了适应智能社会行业用人需求的技能变化，高校需及时调整人才培养目标及培养内容。

另一方面，人工智能正在引发高校师生知识获取、应用、创新方式的巨大变革，对高校教师教学和科研革新提出了时代需求。2024世界慕课与在线教育大会上发布了《世界高等教育数字化发展报告（2024）》和《世界高等教育数字化发展指数（2024）》，并指出数智技术正在引发全球高校全方位变革，全球高等教育进入"智慧教育元年"。

（一）人工智能引发高校师生知识获取方式的变革

知识是人类智慧的结晶。传统高校师生主要通过口耳相授、师生互动、生生互动，以及师生与各种传统媒介（图书、报纸、杂志、电子音像制品，以及互联网）上的内容互动来获取自己想要的知识。大多数知识的获取方式发生在特定的时空内，且师生能接触到的知识量受知识载体限制。通过互联网获取知识的方式虽然在一定程度上能突破传统时空和知识容量的限制，但也会受平台或搜索引擎的质量，以及个人搜索能力的影响。人工智能技术的出现大大拓展了高校师生的知识获取方式，提升了他们获取

---

① UNESCO, "AI Competency Framework for Teachers," 2024, https://unesdoc.unesco.org/ark:/48223/pf0000391104; UNESCO, "AI Competency Framework for Students," 2024, https://unesdoc.unesco.org/ark:/48223/pf0000391105.

知识的效率，智能搜索和智能推荐能够根据师生"教"与"学"的需求，精准定位知识库。

生成式人工智能采用对话式人工智能技术，通过接受超大量的文本训练，结合有监督和无监督的人工智能学习技术，从庞大的数字书、在线文章和其他媒体数据库中获取所需知识，利用这些知识与用户流畅对话，由此大大方便了人类的知识获取。这种便捷的知识获取方式可以助力高校教师更高效地开展教科研工作，方便开发多元化的教育教学资源，包括教案设计、教学案例编写、习题/作业生成；虚拟助教、智能学伴、聊天机器人、智能测评等工具可以帮助教师完成教案生成、课件润色、出题组卷和自动阅卷等教学工作。生成式人工智能可以帮助大学生开展更多对话式学习，完成便捷搜索、快速答疑、个性学习及实时评价等学习活动，助力教与学提质增效。人工智能高校教学应用已经助推教师、学生、人工智能共同参与教学的"师—机—生"新关系的出现。

（二）人工智能引发高校师生知识应用方式的变革

人工智能推动高校师生知识应用方式朝着动态化、智能化和个性化的方向发展，由此推动高校教学与科研的创新发展。在传统教育模式中，知识的应用主要体现在教师授课和学生作业的完成上，教学内容往往是固定且单一的。借助人工智能，教师可以根据学生的学习数据，精准分析其学习进度和知识掌握情况，并实时调整教学内容和方法，助力学生获得个性化学习体验。这种动态的知识应用模式有助于提升教学效率，促进学生自主学习和深度学习的发生。作为高校知识应用的主阵地，高校科研工作也深受人工智能技术应用的影响。在传统科研中，研究者需要耗费大量时间进行数据收集、整理和分析，科研进程往往受到时间和资源的限制。人工智能可以高效处理海量数据，自动生成分析模型，甚至可以提出研究新假设，大大缩短科研周期，为研究者提供新的研究方向，知识应用更加广泛。

人工智能的教育应用促进了信息的自动化收集、知识的跨领域整合、知识呈现的可视化。基于智能技术的个性化推荐和虚拟助手，使得时时、处处、人人可学的泛在学习更加普及，知识应用方式更加个性化，基于真实情境的学习和知识应用愈发普遍。高校从"师—生"二元结构转向"师—机—生"三元结构，推动学习空间泛在化，满足学习过程全覆盖的个性化需求，有助于创建人机协同的学习空间。引入"人在回路"的闭环协同学习机制，形成数据驱动下归纳、知识指导中演绎，以及反馈认知中顿悟等相结合的计算理论模型，将是未来"师—机—生"耦合而成的学习空间中"教"与"学"的发展方向。

新一代人工智能技术具有深度学习、跨界融合、人机协同、群智开放、自主操控等特征，其可解释性、系统偏差、数据安全、数据隐私等问题却给行业和社会带来了前所未有的伦理风险与挑战。如果使用不当，人工智能教育应用也会带来很多负面影响，如教师地位边缘化、学生学习孤岛化、知识体系碎片化、隐私泄露、歧视和偏见、伦理风险、学术诚信和公平失衡、教育关系异化、安全问题、知识盲区与信息茧房、内容准确性不足、学生高阶思维被削弱、数字应用鸿沟等。

（三）人工智能引发高校师生知识创新方式的变革

知识创新是人类文明不断进步的重要基础。知识创新可以表现为对人类未曾涉足的理论或应用领域的发现，对已知现象的科学解释，对已有理论的应用研究，对已知理论体系或应用体系的融合、完善与发展等。传统高校的知识创新通常是由研究者利用传统媒体工具，通过多年的学习积累、科研探索、知识建构和学术交流来完成的。人工智能的出现为知识创新提供了新动力，它可以帮助研究者生成假设、设计实验、计算结果、解释机理，特别是辅助研究者在不同的假设条件下进行大量重复的验证和试错，由此提升科研效率，简化知识创新流程，降低创新门槛，使得高校科研人员从烦琐的实验数据分析及模型构建工作中解放出来，将更多精力用于创意的产生。人机协同模式可以加速科学创新进程，探索先前无法触及的研究领域。

大语言模型以自然语言形式与人类交互，同时将各种应用以插件形式进行整合，成为链姿"人类社会—信息空间—物理世界"三元空间的流量入口。比外，智能体作为能够感知自身环境、自我决策并采取行动的人工智能模型，与生成式人工智能基座模型相结合，形成了人工智能体（AI agents）这一垂直前沿领域。人工智能体在内容合成的基础上，能够实现信息检索、人机对话、任务执行、逻辑推理等自主行为。将人工智能体应用于科研领域，可以提升人类学习能力，拓展未知空间的研究范式。

通过智能算法，特别是 Transformer 这样的神经网络架构，人工智能能够从海量语料中学习单词与单词之间的共生关联关系，实现自然语言的合成。以 Transformer 为核心构建的 ChatGPT 等生成式人工智能系统通过洞悉海量数据中单词—单词、句子—句子等之间的关联性，按照缩放定律（Scaling Law）不断增大模型规模，这一超越了"费曼极限"（Feynman Limit）的模式极大地增强了模型非线性映射能力，具备迅速合成语言的能力，犹如昨日重现一样对单词进行有意义的关联组合，连缀成与场景相关的会意句子，生成有价值的句子和知识。

依靠强大的自然语言处理和生成能力，生成式人工智能不仅可以根据用户的需求和偏好生成个性化内容，还能够理解和处理海量的语料库。这意味着知识生产不再完全依赖人类的个体能力和时间成本，而是可以通过算法实现高效且大规模的生产。这种变革不仅提高了知识生产的效率和速度，还为人类知识的整合、传播和创新提供了全新的可能性，推动高等教育朝着更加智能化和信息化的方向发展。

不过，人工智能生产知识的质量受限于学习算法、训练语料质量、模型规模和提问设计的影响，容易产生过度生成（产生真假难辨、良莠不齐的内容）或生成不足的问题。生成式人工智能的产生使得智能机器成为知识生产的辅助者，对个体学习者的自主思考、判断、学习能力乃至伦理道德观提出了挑战，人类智能有被人工智能边缘化的风险。同时，人工智能驱动的知识生产方式也带来了算法偏见、数据隐私和伦理风险、内容准确性不足、学术不端和公平失衡等问题。因此，高校师生在利用人工智能开展知识生产时需要了解其局限，做到扬长避短，以确保知识生产的公正性和可持续性。

二、智能时代高校教师的定位及技能要求

作为高等教育发展的核心力量，高校教师在人才培养、科研创新、社会服务、文化传承等领域发挥着举足轻重的作用。智能时代，高校教师应准确识变、科学应变、主动求变，重构智能时代各专业的人才培养目标、路径和支持系统，为社会培养大批具有人工智能素养的复合型专业人才，助力未来社会朝着和谐、健康、可持续的方向发展。

（一）智能时代高校教师的角色和定位

高校教师通常有着多年的学术训练，拥有丰富的专业知识储备和较大的科研创新潜力。不同于基础教育阶段以教书育人为主的中小学教师，高校教师通常在高校扮演着更多层次的角色：他们既是大学生成长成才过程中的关键引路人和传道授业解惑者，也是各学科领域的前沿探索者，以及知识生产的积极贡献者，还是社会各行各业实践的重要参与者和变革推动者；同时，很多高校教师还扮演着国家和地方文化传承者与创新者的角色。由此可见，在很大程度上，高校教师的综合素养水平直接影响着高校知

识生产和传播行为、人才培养质量，以及科研创新水平。

虽然智能技术在高校教育教学中能发挥其独特的优势，但高校教师的存在依然是不可替代的。高校教师不仅负责知识的传承和创新，推动科研进步，更重要的是培养综合素质高的未来社会接班人，带领学生利用智能工具改进学习方式、提高学习效率、践行科技伦理、倡导终身学习，更好地服务社会及传承人类文明。

在培养学生核心素养（如人文底蕴、责任担当、国家认同、跨文化交往等），以及发展学生 21 世纪技能（如创造力、批判性思维、复杂问题解决、沟通与协作等）方面，高校教师发挥着人工智能替代不了的作用。不过，要想在智能时代更具胜任力，高校教师需要不断学习，及时更新教育理念，在教书育人和科研创新上掌握新技能。

（二）智能时代高校教师教书育人的技能新要求

在教书育人方面，高校教师应尽快适应人机协同的教育教学环境，熟悉当代大学生的学习偏好和学习风格，及时调整教书育人的目标、内容及手段；熟练掌握学科教学所需的智能化教学工具和平台，更新教学法，更加注重互动式教学和个性化教育，致力于激发学生的学习兴趣和动机，注重培养学生的综合素质和创新能力，同时特别注重学生社会责任感、使命感，以及道德伦理意识的培养。在"师—机—生"构成的高校课堂新生态中，教师应重新思考自身的角色和定位，在传道、授业、解惑中学会充分利用"机师"的优势，同时也尽量发挥人类教师在学生高阶思维培养及人际沟通与协作能力提升方面的独特育人优势。此外，应利用智能技术开展更加科学、高效的教学评价改革，以评促教、以评促学，助力教育目标的实现。

（三）智能时代高校教师科研创新的技能新要求

在科研创新方面，高校教师需要掌握人工智能领域基本的知识和使用技能，从而能够利用各种智能技术使自己和团队更好地保持专业领域的知识更新。有能力和条件的高校教师也可以通过校内外的产学研协同合作，研发学科所需的智能化工具或平台，例如学科领域大语言模型等，更好地开展学科和跨学科领域的科研创新，更科学地确定有意义的研究问题，更全面地收集研究所需的资料和证据，在人机协同的环境下更便捷地推动研究过程的实施，最后，利用智能技术更好地呈现科研成果，实现科研创新的目标。

2

第二部分

高校教师人工智能素养的
概念与内涵

一、高校教师人工智能素养的概念界定

中文"素养"一词出自《后汉书·卷七四下·刘表传》："越有所素养者，使人示之以利，必持众来。"它指的是平日的修养，广义上包含道德品质、外表形象、知识水平与能力等各个方面。在英文中，"素养"通常表达为competence或literacy。

1997年，经济合作与发展组织（OECD）启动了"素养的界定与遴选"（Definition and Selection of Competencies，DeSeCo）项目，围绕"素养"开展了为期九年的研究，最终将素养界定为"在特定情境中，通过利用和调动心理社会资源（包括技能和态度），以满足复杂需要的能力"[1]。同时指出素养具有时代性、整体性、发展性和可测性，即素养依存于特定情境，凡是有助于个体适应社会或解决复杂问题的能力与技巧，都可称为素养（时代性）；素养一般是知识、能力与态度的统一（整体性）；通过特定教育手段，能够实现素养的培养（发展性）；素养能够通过可理

---

[1] OECD, "The Definition and Selection of Key Competencies: Executive Summary," 2005, https://www.deseco.ch/bfs/deseco/en/index02.parsys.43469.downloadList.2296.DownloadFile.tmp/2005.dskcexecutivesummary.en.pdf.

解、可操作、可评估的指标进行度量（可测性）。

（一）教师素养的概念及内涵

《中华人民共和国教师法》规定了教师所应具备的基本要素。教师首先应当遵守法律和职业道德，成为学生的良师益友。教师还应该在完成教学工作任务的同时，全心关注、照顾每位学生，尊重学生的人格，促进学生全面发展。此外，教师应该提高自身的思想和教学水平，提升专业技能和素养水平，从而更好地履行育人使命。教师素养的内涵随着社会的不断发展也在不断丰富。[①]新时代教师素养的要求涵盖多个方面，包括思想政治素养、文化素养、专业素养、现代素养等。思想政治素养是教师必备的素养之一，即具备正确的历史观、政治观、价值观、人生观等方面的素质。文化素养则要求教师有广泛的知识储备和深厚的文化底蕴，能够领略文化的魅力，并把这些知识和文化传授给学生。专业素养要求教师具备扎实的学科知识和教育教学理论知识，能够灵活运用各种教学资源和手段，以达到有效促进学生学习的目的。现代素养则要求教师能够熟练掌握现代化的教育技术手段，善于创新和应用，实现优质教育。[②]

（二）人工智能素养概念的提出

随着人类进入智能时代，人工智能素养（AI literacy）逐渐

---

① 中华人民共和国教育部：《中华人民共和国教师法》，2009年，http://www.moe.gov.cn/jyb_sjzl/sjzl_zcfg/zcfg_jyfl/ tnull_1314.html。
② 许扬：《新时代高校思想政治理论课教师素养提升研究》，广西师范大学硕士学位论文，2023年。

成为个体生存和发展的重要素养之一。这一概念的首次提出是在 20 世纪 70 年代，当时主要强调的是人工智能专业技术人员的素养组成。[①] 2017 年以来，人工智能对人类社会的影响越来越大，提升公民人工智能素养的议题日益受到社会各界关注。国内外多位学者对人工智能素养的概念进行了界定，例如，学者朗和马杰可（Long & Magerko）指出，人工智能素养源自并扩展自技术相关素养（信息素养、数字素养、计算素养、数据素养等），它包括"个人能够批判性地评估人工智能技术、与人工智能有效沟通与协作，并在在线、居家和工作场所使用人工智能所需的系列技能"[②]。学者黄家伟（Wong）等认为人工智能素养包括"AI 概念、AI 应用、AI 伦理/安全"三个部分，其中，"AI 概念"是指基本的人工智能知识与起源，"AI 应用"是指人工智能技术在现实世界中的应用，"AI 伦理/安全"是指人工智能应用过程中所面临的道德挑战和安全问题。[③]

2021 年 12 月，UNESCO 在"教育中的人工智能"主题会议上提到人工智能素养的重要性，并从广义上将人工智能素养界定为关于人工智能的知识、理解、技能和价值取向，认为所有公民都需要具备人工智能素养——这已成为 21 世纪的基本语法。

[①] Agre, P. E., "What to Read: A Biased Guide to AI Literacy for the Beginner," MIT Artificial Intelligence Laboratory Working Papers, WP-239, 1972.

[②] Long, D., Magerko, B., "What Is AI Literacy? Competencies and Design Considerations," Proceedings of the 2020 CHI Conference on Human Factors in Computing Systems, 2020, pp. 1-16.

[③] Wong, G. K. W., Ma, X., Dillenbourg, P. et al., "Broadening Artificial Intelligence Education in K-12: Where to Start?," ACM Inroads, 2020, 11(1): 20-29.

2024年，UNESCO发布了《面向教师的人工智能能力框架》，其中指出，智能时代的教师应该在五个维度拥有人工智能相关的能力，即以人为本的思维、人工智能伦理、人工智能基础与应用、人工智能教学法，以及人工智能与专业发展。[①]该框架为全球各级各类教师的人工智能素养的提升指出了方向。

（三）高校教师人工智能素养概念的界定

身处智能时代，高校教师教书育人、科研创新、社会服务、文化传承等活动都面临新的机遇和挑战。在最主要的教育教学和科研活动中，随着人工智能的介入，其教学和科研的目标、内容、方法、手段、评价等维度都将发生根本性颠覆和革新。高校教师需要直面这些重大变化，及时更新自身的教育理念，掌握和更新与人工智能相关的知识和技能，借人工智能之力，提升自身的创新能力，同时对人工智能持有正确的价值观和伦理原则。

鉴于高校教师群体的特殊性和重要性，探讨高校教师人工智能素养的概念、内涵和提升对于智能时代高校变革具有重要意义。在借鉴国内外已有人工智能素养/能力的概念及内容框架的基础上，结合高校教师的工作特点，本书将高校教师人工智能素养界定为在高校从事教学与科研工作的教师为了在智能时代胜任教书育人、科研创新、社会服务和文化传承等工作而应具有的与人工智能应用相关的专门素养。它包含赓续育人理念（何为师）、

---

① UNESCO, "AI Competency Framework for Teachers," 2024, https://unesdoc.unesco.org/ark:/48223/pf0000391104.

掌握智能知识（以何为师）、变革教研模式（何以成师）和担当社会责任（师者何为）等能力。具体而言，高校教师人工智能素养包括智能时代育人理念、智能教育基本知识、人机协同教学能力、数智赋能科研创新和科技向善人本价值等五个维度的内容。其中，理念引领、知识为基、能力为核、创新为重、价值为本，五者相辅相成、相互融合（图1）。

图1　高校教师人工智能素养的概念

二、高校教师人工智能素养的内涵解读

高校教师人工智能素养的内涵包括以下要点。

（一）智能时代育人理念

智能时代，人类知识生产和传播的模式正在发生巨大变化，人机协同/互动成为师生学习和工作的常态，各种教育教学资源唾手可得，智能推荐使得高校育人模式和师生关系发生根本改变，各种良莠不齐的信息也因此更有机会影响高校教师和大学生的人生观、价值观和世界观。因此，高校教师需要及时更新育人理念，除了继续以"德高为师、身正为范"标准要求自己成为新时代合格教师之外，还应将言传身教融入教书育人全过程，更加注重智能时代大学生的品德培养和人格塑造，以"发展全人"为人才培养目标，通过技术赋能育人过程，重构育人场景和行为。这一新理念可以表述为：立德树人是根本，创新思维是目标，因材施教是关键，理实结合是路径，技术赋能是保障。表1呈现了"智能时代育人理念"维度的内容。

表1 "智能时代育人理念"维度的内容描述

一级维度	二级维度	描 述
智能时代育人理念	立德树人是根本	理解智能时代高校人才培养中立德树人的重要意义
	创新思维是目标	明确智能时代高校人才培养中发展学生创新思维是重要目标
	因材施教是关键	知晓智能时代开展个性化教育和因材施教是高校人才培养改革的关键
	理实结合是路径	倡导和践行智能时代人才培养过程中理论与实践的紧密结合
	技术赋能是保障	认同并接受智能时代技术在高校教学和科研等活动中发挥的积极保障作用

具体而言，"智能时代育人理念"维度希望高校教师能够深

刻理解"培养德智体美劳全面发展的社会主义建设者和接班人"是中国高校人才培养的根本任务，知晓智能时代高校德育面临的新挑战，以及立德树人对于个人、家庭、社会和国家的重要意义；明确智能技术能够赋能人类的生产、生活和学习，创新思维是智能时代全球公认的人才核心竞争力所在，因此，发展大学生创新思维是高校育人的重要目标；知晓智能技术使得个性化教育和因材施教成为可能，并且是高校人才培养改革的关键；理解智能时代理实结合对于高校创新人才培养和知识发现的重要性，在教育教学和科研活动中倡导并践行理论与实践的紧密结合；了解智能技术在高校教学和科研等活动中的独特优势，认同并接受其发挥的积极保障作用。

（二）智能教育基本知识

智能教育旨在利用智能技术，优化学科教育教学过程，推动学科教学方法改革，以及人才培养模式变革，尝试建立以学习者为中心的教育环境，提供学科内容的精准推送服务，实现学科日常教育和终身教育定制化的教改目标，其核心在于通过智能化手段改进教学方法、管理和评估，以适应不同学习者的学习需求和能力水平。高校教师需掌握智能教育基本知识，包括智能技术基本原理、智能教育理论基础、智能教学工具/平台、智能教学与评价方法、智能教育管理知识等。表2呈现了"智能教育基本知识"维度的内容。

表2 "智能教育基本知识"维度的内容描述

一级维度	二级维度	描 述
智能教育 基本知识	智能技术基本原理	了解学科领域涉及的人工智能基础性知识及人工智能技术的基本原理
	智能教育理论基础	熟悉国内外智能技术教育应用的概况、趋势及相关理论基础
	智能教学工具 / 平台	熟悉学科领域常见的智能教学工具 / 平台及教学应用的案例
	智能教学与评价方法	掌握利用智能技术开展学科教学及评价的典型方法
	智能教育管理知识	掌握利用智能技术开展学科教育教学管理的基本知识

具体而言,"智能教育基本知识"维度希望高校教师能够对智能技术的基本原理,尤其是与其学科领域相关的人工智能基础性知识与技术原理,有初步了解,如有关数据、算法和算力的基本知识,有关图像识别、语音识别与合成的基本知识,有关大语言模型及智能体的技术原理等,对该学科领域中智能教育技术的使用知根知情;能够熟悉支持智能技术应用于其学科领域教学科研的理论,如整合技术的学科知识框架(TPACK)、多媒体认知理论、生成学习理论、联通主义学习理论等,能基于理论分析国内外智能教育的成功案例;能够熟悉所在学科领域具有代表性的智能教学工具和平台,了解国内外高校使用智能教学工具的经典案例,为学科领域的智能教学积累工具资源;在开展智能教学活动时,能够根据学习者特点及课程内容特性,选择合宜的智能

技术开展学科教学及评价，提升自身教学效率，减轻评价工作负担。此外，高校教师需要掌握学校常用的教育教学管理系统（教务管理系统、科研管理系统等）中智能技术的使用原理及规范。

（三）人机协同教学能力

高校教师在智能教育中采用人机协同开展教学活动旨在在学科教学领域践行"以生为本"理念，通过利用人工智能技术，更好地了解学习者的初始水平和学习需求，确定合理的学习目标；选择合宜的教学方法，设计智能化教学环境和学习体验；使用多元化评价方法，针对学习者开展以增值性评价为主的学习评价。为了更好地开展人机协同教学，高校教师需要努力提升自身的设计思维水平，以学习者为中心，强调对学习者需求和教学问题的深刻理解，通过迭代的设计过程，生成富有创意且实际有效的解决方案。设计思维要特别关注同理心、定义问题、构思、原型设计、测试及迭代等要素。表 3 呈现了"人机协同教学能力"维度的内容。

表 3　"人机协同教学能力"维度的内容描述

一级维度	二级维度	描　述
人机协同教学能力	确定合理的学习目标	能够利用一定的智能技术，更好地了解学习者的初始水平和学习需求，确定合理的多维学习目标
	选择合宜的教学方法	能够利用一定的智能技术，更好地确定学习者及学习内容特征，据此选择合宜的教学方法

续表

一级维度	二级维度	描　述
人机协同教学能力	设计智能化教学环境	能够利用一定的智能技术，设计有利于学习目标达成和教学方法实施的智能化教学环境
	设计学习者学习体验	能够利用一定的智能技术，设计有利于学习目标达成的学习者学习体验活动
	设计多元化学习评价	能够利用一定的智能技术，设计有利于检测学习目标是否达成的多元化学习评价内容

具体而言，"人机协同教学能力"维度希望高校教师能够运用智能技术优化课堂教学的理念，在教学实施前对学习者的知识、技能等初始水平及学习需求进行动态了解，确定合理的多维学习目标，促进学习者智慧与能力相长；能够依托智能技术，精准了解学习者及学习内容的典型特征，发现学习者的盲点及偏好，选择合宜的教学方法；能够借助智能技术，以及交互、感知等设备，创建虚拟、真实、多样的教学环境，设计有利于学习目标达成和教学方法实施的智能化教学环境；能够利用智能技术促进教学过程中学习者学习任务的执行能力，为学习者设计情境化的学习感知与沉浸式的具身体验，为更好的知识理解与探究提供学习体验活动；能够充分发挥人机优势，融合智能评价与教师评价，设计多元化学习评价内容，支持对学习者的综合性评价、增值性评价，给予学习者动态、精准、科学的评价反馈，实现因材施教。

（四）数智赋能科研创新

通过人工智能与科学挑战和工程难题等应用场景及任务的结合，将人工智能算法架构在人类不同学科专业知识之上，以人机协同方式促进教育创新、科学发现和工程突破，并进一步推动智能系统的探索和实现。智能化科研（AI for Research，AI4R）这一范式正深刻影响中国科技发展前途及高校教师科研创新实践。

与传统科研创新不同，智能时代科研创新具有不确定性、复杂性和学科交叉性，高校教师需要训练在智能化科研环境下的创新力，包括在优化科研知识结构、确立明确的科研方向、引入和运用最新科研方法、科研过程执行和研究发现描述等关键科研环节中采取基于人工智能技术的科研创新行动。同时，在条件许可的情况下，高校教师需要参与学科领域相关的人工智能研究与开发，推动智能系统自身的迭代提升。表4呈现了"数智赋能科研创新"维度的内容。

表4 "数智赋能科研创新"维度的内容描述

一级维度	二级维度	描述
数智赋能科研创新	创新科研知识的获取	能够利用数智技术创新科研知识的获取途径或方法
	创新研究问题的选择	能够利用数智技术创新研究问题的提出及选择方式
	创新研究方法的使用	能够利用数智技术创新研究方法的选择及使用，包括利用大数据的方法及交叉学科的方法

续表

一级维度	二级维度	描　述
数智赋能科研创新	创新研究过程的开展	能够利用数智技术创新研究的开展过程，人机协同是其基本特征
	创新研究结果的呈现	能够利用数智技术创新研究结果的获得及呈现

具体而言，"数智赋能科研创新"维度希望高校教师能够借助数智技术，从史无前例增长的科研数据中快速筛选、分析和定位有价值信息，发现隐藏的研究趋势和关联，创新科研知识获取的途径或方法；能够依托数智技术对人类历史长河中点滴碎片化研究进行整体建模，开启研究者的直觉顿悟，寻找未完全解决或者尚待解决的前沿挑战，增强人类创造力；能够通过数智技术模拟任意条件下的无穷尽实验过程，减少实验成本和时间开销，探索设计新的实验过程及分析手段，让人工智能成为多学科交叉的黏合剂和催化剂，帮助高校教师创新研究方法；通过人机协同工作，人类与机器在特定任务中相互配合，以实现更高的工作效率和创造力，将人类独有的推理优势与机器的高效率搜索完美融合，从而解锁解决复杂问题的巨大潜力；在创新研究结果呈现方面，高校教师需要利用数智技术创新结果可视化形式，使研究成果更直观、生动地呈现，突破费曼极限，增强学术交流的效果和公众理解，提升研究的传播力和影响力。

（五）科技向善人本价值

《礼记》有云："师也者，教之以事而喻诸德也。"高校教师不仅以事育人，更以德育人。在智能教育过程中，高校教师肩负教学、科研双重使命，其科技向善的人本价值与道德理念对学生学习生活和科研工作具有深远影响。当下，人类已进入以"信息空间—物理世界—人类社会"三元空间为特征的新生态环境之中，其中的伦理价值议题不仅涉及人与人之间的关系和人与自然界既定事实之间的关系，还涉及人类与人造物在社会中所构成的关联，使得人工智能具有技术和社会双重属性。因此，在人机共融社会中，人类应遵守科技向善、以人为本的价值理念，确保把人类价值观、道德观和法律法规贯穿于人工智能的产品和服务，赋予人工智能社会属性。表5呈现了"科技向善人本价值"维度的内容。

表5 "科技向善人本价值"维度的内容描述

一级维度	二级维度	描　述
科技向善人本价值	数据安全与隐私保护的意识	在使用智能技术过程中能够有意识地注意到数据安全问题与隐私保护的重要性
	算法偏差与模型幻觉的警惕	在使用智能技术过程中能够知晓算法偏差与模型幻觉是如何产生的及其后果
	科技向善和以人为本的对齐	在使用智能技术过程中能够始终坚持遵守科技伦理和学术诚信原则，坚持科技向善的目的及以人为本的基本原则
	人机共生共融和全民普及理念	在使用智能技术过程中能够坚持人机共生共融，以及全民普及的理念
	人类累积知识普惠共享的追求	能够坚持追求利用智能技术促进人类知识的累积，以及面向全人类的普惠共享

　　具体而言，"科技向善人本价值"维度希望高校教师能够在智能技术与教育深度融合的环境中主动反思并理性审视智能技术可能带来的优缺点，坚守科技进步为人类福祉提升服务的宗旨，警惕生命物化导致人类被摧毁这一悲剧发生的可能性。在人机协同过程中，能够有意识地关注师生个人隐私数据，以及科研隐私数据在收集、传输、存储、备份及分析时的安全性；能够了解并深刻理解算法模型的自身特性及其影响，包括不可解释性、有偏性、幻觉及谄媚性等的表现及后果，自身在使用时保持底线思维和风险意识，同时引导学生理性消费技术；在开展教学和科研过程中，能够始终自觉坚持并引导学生遵守科技伦理原则，建立科技向善和以人为本的基本原则与价值观；能够利用人工智能所具有的社会属性，确保科技活动中使用人工智能风险可控，科技成果造福于民，实现人工智能全民普及理念；能够积极思考如何利用智能技术促进本专业领域知识的累积，将面向全人类普惠共享学科前沿知识作为追求。

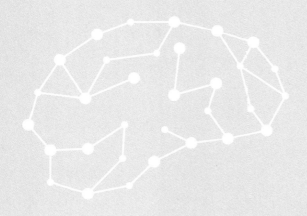

3

第三部分

高校教师人工智能素养提升的
目标、路径与保障

一、高校教师人工智能素养提升的目标

高校教师人工智能素养提升的目标是：经过多形式的学习、交流和实践活动，高校教师能够具有智能时代先进的育人理念、掌握智能教育领域的基本知识、具备人机协同环境下开展教学的能力、拥有数智技术辅助的科研创新能力，以及坚持科技向善和以人为本的价值取向。

二、高校教师人工智能素养提升的路径

为贯彻落实《中共中央 国务院关于全面深化新时代教师队伍建设改革的意见》，深入推进人工智能等新技术与教师队伍建设的融合，推动教师主动适应智能时代教育变革的趋势，教育部于 2018 年在宁夏回族自治区和北京外国语大学启动了人工智能助推教师队伍建设试点工作，2021 年进一步扩大试点规模，在 56 所高校、20 个地市，以及 25 个区县开展了更多内容和更深层次的建设应用。国际上，UNESCO 近年连续发布了几份有影响力的全球报告，包括《高等教育中的 ChatGPT 和人工智能：快速入门指南》（2023）、《教育和研究中生成式人工智能的使用指南》（2024）和《面向教师的人工智能能力框架》（2024）等，旨在引导高校教师更好地将人工智能融入各国高校。

由此可见，聚焦高校教师人工智能素养提升的薄弱环节，重点推进教师应用智能助手（平台、系统、资源、工具等），促进教科研方式和学习方式改革，为教师减负和赋能，这是全球公认的应对智能时代高校变革的重要举措。高校教师人工智能素养提升的主要路径包括以下内容。

（一）更新高校教师培训的内容及形式

随着人工智能在高校教育教学和科研活动中应用的不断深化，高校人事管理部门，以及负责教师专业发展的部门有必要围绕"AI＋教育"专题有组织地开展面向不同教师群体（新入职教师、有一定教科研经验的中青年教师、有多年教科研经验的资深教师）的在职培训，重点提升高校教师利用智能技术开展教育教学及科研创新实践的水平。为提升高校教师人工智能素养五大维度的内容，学校可开展以下活动。

（1）通过学校和学院层面的教师大会、双代会、教学研讨会、新老教师培训、导师学校、务虚会等活动，宣传智能时代育人理念的新内涵，分享代表性高校和学科的先进育人理念案例，不断提升全体高校教师智能时代育人理念。

（2）通过邀请智能教育领域专家开展主题报告，向各学科教师普及人工智能的发展史及最新进展、人工智能的基础知识和使用技能、智能教育基本知识。有条件的学校可以围绕智能教育基本知识制作相关慕课或微课资源，放在学校相关的学习平台上，供高校教师自学。教师课程学习的成果可被纳入学校人事管理或教师发展中心等部门设立的教师培训学分认定系统。

（3）本科生和研究生教学管理部门可组织高校教师系统学习国内外有代表性的人工智能高校教学应用指南，了解人机协同教学实践的实施路径、注意事项和全球最佳实践案例，由此提升教师人机协同教学能力。有条件的学校可以针对本校学科特色和师生特点，尝试制定人工智能教育教学应用指南，由此指导和规范教师人机协同教学实践。教师发展部门可以通过创新教学工作坊、教改项目申报及经验分享等活动，邀请高校教师亲身实践人机协同教学过程。

（4）高校科研管理部门、信息技术中心（或学校相关技术支持部门）和教师培训部门一起收集并整理全球范围内智能技术辅助科研创新的典型案例，通过学校、学院及学科层面的各类教师教科研培训、项目申报动员会、项目申请经验分享会、研究生教育专题会等活动，分享给各院系教师。通过这些活动，高校教师能够较快地熟悉领域内能够助力科研活动的智能技术及使用方法。此外，工作坊、实地参观考察等活动有助于让高校教师亲身体验智能技术辅助科研创新过程。

（5）学校教学管理部门、科研管理部门，以及教师培训部门可以邀请科技哲学、科技伦理、人工智能伦理等交叉学科领域专家，为不同学科教师介绍科技伦理和人工智能伦理领域的基本知识、法律法规，以及最佳实践做法，通过讲座、案例研讨、主题辩论、工作坊等形式不断启发不同学科教师思考智能时代科技向善及人本伦理的概念、内涵与规范做法，由此引导高校教师从思想上到行为上，努力遵循科技向善和人文价值的技术使用观念。

围绕高校教师人工智能素养提升的教师培训形式应尽量多样化，以满足教师培训的不同目的和需求，既可以包括主要用于人工智能知识和教学经验传授的传统讲座，又可以包括发展人工智能应用技能的工作坊、实践活动或短期课程，还可以包括启发式、对话式的研讨和辩论活动。培训开展形式可以是面对面的，也可以是线上线下混合的，抑或是完全在线的形式。

（二）创新高校教研室的活动内容与形式

高校教研活动是促进教师专业发展、提升教师教学能力的重要途径，高校教研室是组织教师开展教研活动的重要部门。高校教研室起源于 20 世纪 50 年代，是按专业或课程设置的基本教学和研究单位，是致力于高校教育教学改革和教师专业发展的重要基层教学组织，在高校新老教师的"传帮带"过程中，以及提高教师专业素养方面发挥着重要作用。传统高校教研室的活动通常在指定时间和地点开展，教师面对面研讨教育教学中出现的问题，以及分享各种经验，其目的是促进更多教师提升教育教学质量。为更好地顺应智能时代的高等教育变革趋势，高校教研活动的内容和形式有必要推陈出新。在教研活动内容方面，鉴于人工智能对高校日常教与学的行为正在产生日益深刻的影响，高校和院系层面的教研活动有必要有组织、有目的地将此话题作为重要议题开展持续性研讨，从教师教学、学生学习和学校教学管理三大维度研讨如何更安全、有效地将人工智能融入高校教学活动。

在教研活动形式方面，除了传统面对面的教研活动，各高校可以充分利用信息技术优势开展基于网络的教研活动。教育部于2021年发布了《关于开展虚拟教研室试点建设工作的通知》，提出了开展高校虚拟教研室建设的行动计划，并于2022年先后批准了657个高校虚拟教研室试点，鼓励专业建设类、课程（群）教学类，以及教学研究改革专题类三种类型的虚拟教研室建设。虚拟教研室作为一种创新的教学组织形式，可以突破时空限制，整合校内外优质教学资源，构建良好的教学科研生态，是促进高校教育教学改革的重要基层教学组织。作为基于现代信息技术构建的新型基层教学组织，虚拟教研室能够有效整合优质教学资源，实现知识融合与创新，构建灵活的师资队伍，满足交叉专业的教学需求；其开放性能够突破传统时空和学科限制，促进不同院校协同教学、多学科融合、资源共享，拓宽专业视野。这些优势使虚拟教研室可以解决高校跨学科师资配置和科研团队知识结构的优化等问题，使教研工作更加动态、开放，不受时间、空间和地域限制。它是智能时代高校推动跨学科专业建设、培养复合型人才的重要力量。附录1呈现了教育部批准的与人工智能有关的高校虚拟教研室名单及案例。

（三）项目驱动锻炼高校教师融合人工智能的实践能力

国内外经验表明，基于项目的学习是学习者实践能力提升的有效途径。高校教师通过主持或参与 AI ＋ X 和 X ＋ AI 主题的各类教改或科研项目，可有针对性地开展融合人工智能的专业教学实践及创新科研路径探索，提升跨学科的实践能力。

教育部高教司、各省份高教主管部门，以及各级各类高校的本科生教育和研究生教育管理部门可以通过设置 AI ＋ X 和 X ＋ AI 的专项教改项目，鼓励高校各学科教师参与人工智能融入的教育教学改革实践和研究。通过这些教改项目，高校各学科教师可以积极探索人工智能教学应用的可能性，比如使用人工智能工具更好地开展教学设计或进行学习者学习过程及结果的分析，师生一起探讨人工智能在教育教学应用中的优缺点。通过学科教学的实际案例分析，帮助教师了解人工智能在教育中的成功应用或潜在挑战，培养高校教师在智能环境下优化教学方案和解决棘手教学问题的能力。基于教改项目的科研成果（论文、著作、发明、专利等），也有利于同行交流和创新扩散。

此外，教育部、科技部，以及各省份科研管理部门可设立 AI ＋ X 和 X ＋ AI 主题研究项目，鼓励不同学科的高校教师将人工智能技术应用于学科科研活动，并鼓励非计算机学科的教师与计算机科学、人工智能、数据科学等专业的同行，以及企事业单位开展跨学科、跨部门的合作研究，分享彼此的知识和经验，来自不同学科背景的高校教师取长补短，最终促成人工智能较好地融入高校各学科的科研创新活动。

三、高校教师人工智能素养提升的保障

高校教师人工智能素养的提升需要高校在组织、制度、资源和环境等四大方面提供强有力的保障。

（一）组织保障

各级各类高校领导班子首先需要充分意识到智能时代高等教育变革的紧迫性，以及高校教师人工智能素养提升的必要性。在此基础上，学校领导班子需要围绕如何积极、有效地推进高校教师人工智能素养提升工作开展顶层设计及相关组织机构的配置，如成立专门工作指导小组和/或人工智能教育教学研究中心来确保该项任务后续的有序推进。这些组织成立的主要目的是推进高校教师人工智能素养的相关培训、实践和研究，通过研制相应的政策、报告或指南，在全校范围内引导教师更好地开展人工智能相关的教学和科研活动。此外，学校人事主管部门、本科生管理部门、研究生管理部门、教师专业发展中心、继续教育培训部门和各院系等多个组织应更好地开展跨部门协同，通过调动校内外尽可能多的资源，为各发展阶段的教师（新教师、有一定教研经验的中青年教师及资深教师）人工智能素养的提升提供强有力的组织保障。

（二）制度保障

各级各类高校可以通过体制机制创新为高校教师人工智能素养提升提供良好的制度保障。一是将教师培训、教研活动和教改

项目的参与情况以及教学成果奖、教学技能比赛等纳入高校教师晋升及评聘条件，在全校范围内提升教师参与智能教育和智能化科研活动的积极性与主动性。二是参照国内外已有的一些指南（附录2），制定人工智能（尤其是新一代生成式人工智能）融入高校教育教学和科研的师生使用指南，以此规范师生使用行为。UNESCO 于 2023 年 4 月发布的《高等教育中的 ChatGPT 和人工智能：快速入门指南》（附录3），以及国内外很多高校已经发布的相关指南可以成为广大高校相关制度建设的重要参考。

（三）资源保障

各级各类高校可以从人力、物力和财力等方面为高校教师人工智能素养提升提供良好的资源保障。具体而言，首先，为高校师生提供免费或低费用的人工智能融入的教学或科研工具/平台，尤其是生成式人工智能的访问权限。这类资源可以从校外引进或由高校自主研发。本科生和研究生教务管理部门、学校信息技术支撑部门等部门联动，为教师积极参与研发学科类、专业类人工智能助手，以及参与学科垂直领域大语言模型研发提供必要的资源支持。其次，高校在教师培训、教研室活动、教改项目等方面保证一定的人力、物力和财力投入，尝试为教师提供定制的 AI 素养提升路径，吸引更多高校教师参与和推广智能教育与智能化科研，鼓励教师开展跨学科交叉和融合实践。

（四）环境保障

各级各类高校可以从物化及文化两个层面为高校教师人工智

能素养提升提供良好的环境保障。一方面，高校有必要根据自身学校和学科特点，选择适切的数智技术，为本校师生构建一个有利于开展智能教育和智能化科研的数字化环境，包括数智技术支持的物理空间（智慧教室、创客空间、AI实验室、图书馆等）和数字空间（教学、科研、管理系统/平台及数字孪生校园等），使师生有机会在智能的物理空间及数字空间开展创新的教育和科研实践。另一方面，高校及相关企事业单位可以为师生创设良好的人机协同育人和协同科研的文化环境。这种环境鼓励高校开展产学研合作，鼓励各专业师生开展跨学科交流与合作、跨文化交流与合作，鼓励各种形式的创新思维和创新实践，鼓励各种数智技术的研发与应用，由此重构传统高校新生态，为智能时代高校变革提供充分的环境支持。

结　语

　　以 ChatGPT 为代表的生成式人工智能会深刻影响人类社会的未来发展，为此，全球高等教育都在进行积极变革，以应对智能时代带来的机遇和挑战。作为国家各行各业大批量创新人才培养最重要的阵地，中国高校肩负着实现国家教育科技人才强国的光荣使命。无论技术环境如何改变，教育的最终目的是促进人的发展，也因此，智能技术融入高等教育的终极目标也是要促进人的全面发展。

　　为了顺应国际大趋势，我国高校正在积极实施国家教育数字化战略行动计划，加快推进高等教育数字化转型、智能升级及融合创新，支撑高等教育高质量发展。其中，全方位提升高校教师队伍素质是关键，尤其是高校教师有关人工智能的知识和技能，教师的使用行为及态度将决定高校智能教育是否发生、如何发生，以及发生后的效果和质量。

　　本红皮书发布的初衷是提醒广大高校教师及时更新自己的"装备"，以此应对"第四次工业革命"带来的史无前例的冲击。广大高校教师需要对时代的巨变有深刻的认识，并意识到自身需要不断学习，从多个维度提升人工智能素养，赓续育人理念、掌握智能知识、变革教研模式和担当社会责任，由此确保自己能够更好地胜任高校教师这一职业，为未来社会培养更多有竞争力的创新人才，同时通过科研创新为人类知识生产和文化传承更好地贡献自己的力量。

附　录

附录1：与人工智能有关的高校虚拟教研室名单及案例

在教育部2022年首批和第二批虚拟教研室试点名单中，新工科类占比较高，约40%，有17个与AI相关的项目（如附表1所示）。

附表1　教育部首批及第二批与人工智能有关的虚拟教研室试点名单

类　型	虚拟教研室名称	学校名称	带头人
专业类	音乐专业（音乐人工智能方向）虚拟教研室	中央音乐学院	俞　峰
	城乡规划专业（智能城市与智能规划方向）虚拟教研室	同济大学	吴志强
	人工智能专业（AI+X方向）虚拟教研室	浙江大学	吴　飞
	建筑电气与智能化专业虚拟教研室	安徽建筑大学	方潜生
	智慧牧业科学与工程专业虚拟教研室	西北农林科技大学	姚军虎
	车辆工程专业（轨道车辆智能运维方向）虚拟教研室	北京交通大学	刘志明

类　型	虚拟教研室名称	学校名称	带头人
课程（群）类	"101 计划"人工智能引论课程虚拟教研室	浙江大学	吴　飞
	人工智能课程虚拟教研室	浙江工业大学	王万良
	道路工程智能建养课程群虚拟教研室	长沙理工大学	袁剑波
	建筑智能化实验课程群虚拟教研室	西安建筑科技大学	于军琪
	智能＋新农科课程虚拟教研室	西北农林科技大学	李书琴
	智海 AI 课程虚拟教研室	哈尔滨工程大学	刘海波
	智慧会计课程群虚拟教研室	东南大学	陈志斌
	智能制造课程群虚拟教研室	宁夏理工学院	巩云鹏
教改类	智能时代下体育与健康教学研究虚拟教研室	华东师范大学	汪晓赞
	智能林业装备人才培养模式研究虚拟教研室	南京林业大学	周宏平
	智慧林业人才培养模式改革虚拟教研室	南京林业大学	曹福亮

作为教育部首批虚拟教研室建设单位之一，浙江大学人工智能专业（AI＋X方向）虚拟教研室依托于计算机学院 2020 年创建的AI＋X微专业教研中心。该虚拟教研室创设了"教材建设、课程共享、平台增效"三位一体的建设模式。其中，教材建设引领课程和平台在育人理念、培养模式与教学内容上的总体方向；课

程共享是围绕教材建设的宗旨，充分发挥各方智慧、解决各方问题的智库，为平台功能建设提供架构设计；平台增效是检测教材和课程的实训工具，推动"知识本位"向"能力本位"转变，实现知行合一。三者相辅相成，突破传统专业界限，探索学科交叉人才培养新模式，实现知识的有效融合。其建设思路具体包括以下几点。

依托 AI+X 微专业建设，拓展产教协同人才培养途径。该虚拟教研室通过与产业集群合作共建 AI+X 微专业，依托产业集群中各细分领域的实际需求，为非计算机专业学生提供学习人工智能核心理论和实践应用的机会，培养具备复合知识结构和实践能力的交叉型人才。通过微认证（Micro-credential）项目，采用应用导向和问题导向的培养方式，将产业实践与教学深度融合，培养能够适应新技术、新业态发展需求的应用型人才。

建设 AI+X 优质共享资源，推动区域资源共享。该虚拟教研室通过校际合作，系统规划并编写系列 AI+X 教材。教材的内容创新点体现在：一方面，突出交叉领域基础理论和前沿应用知识，并辅助开发相应配套视频。另一方面，制定资源开发标准和共享机制，推动内容动态更新迭代，鼓励适宜的教学内容向全国高校开放共享，为 AI+X 人才培养提供更为可持续、开放的支持。

优化 AI+X 科教创新平台，打造开放创新社区。该虚拟教研室以智能科教与实训平台为依托，开源开放的算法、模型和数据等资源，提供丰富的数字化学习资源和工具，引导并提升学生的实践实训应用能力。基于 AI+X 科教创新平台，教研室构建领域知识库、计算资源池和能力测试标准，打造跨学科、开放融合的创新社区。这一社区不仅能够激发师生的探索兴趣，而且有助于实现理论知识、课程实践、行业解决方案和产品落地的全流程打通。

加强AI+X师资培训，提升师资教学能力。该虚拟教研室开展以教师成长为核心的系统性的AI+X培训体系，除了学科领域的前沿内容以外，还包括教学方法、教学技术等方面的培训，以全面提升教师的教科研能力。虚拟教研室创新"问题+方案"的教研形态，打破传统的学科界限，探索学科交叉汇聚的教研模式，以问题解决为驱动，促进不同学科教师之间的交流与合作，全面提高教师问题导向意识和协作解决教学环节中复杂问题的能力。

附录2：国内外代表性高校生成式人工智能使用指南名录

国内外代表性高校生成式人工智能使用指南名称及网址见附表2。

附表2　国内外代表性高校生成式人工智能使用指南名称及网址

高　校	名　称	网　址
哈佛大学文理学院	哈佛大学关于使用ChatGPT及其他生成式人工智能工具的指南	https://huit.harvard.edu/ai/guidelines
耶鲁大学	生成式人工智能工具使用指南	https://provost.yale.edu/news/guidelines-use-generative-ai-tools
斯坦福大学	生成式人工智能政策指南	https://communitystandards.stanford.edu/generative-ai-policy-guidance
麻省理工学院	用ChatGPT教与学：机遇还是困境？第三部分	https://tll.mit.edu/teaching-learning-with-chatgpt-opportunity-or-quagmire-part-iii/
牛津大学	ChatGPT的四点启示：教育工作者面临的挑战和机遇	https://www.ctl.ox.ac.uk/article/four-lessons-from-chatgpt-challenges-and-opportunities-for-educators
剑桥大学	我们如何使用生成式人工智能工具	https://www.communications.cam.ac.uk/generative-ai-tool-guidelines
卡耐基梅隆大学	生成式人工智能工具常见问题解答	https://www.cmu.edu/teaching/technology/aitools/index.html
加利福尼亚大学伯克利分校	了解人工智能写作工具及其在加利福尼亚大学伯克利分校教学与学习中的应用	https://teaching.berkeley.edu/understanding-ai-writing-tools-and-their-uses-teaching-and-learning-uc-berkeley

高　校	名　称	网　址
加利福尼亚大学洛杉矶分校	加利福尼亚大学洛杉矶分校（UCLA）的生成式人工智能在教学中的应用	https://teaching.ucla.edu/resources/teaching-bulletins/
南加利福尼亚大学	学生在学术作业中使用生成式人工智能的教师指南	https://academicsenate.usc.edu/wp-content/uploads/sites/6/2023/02/CIS-Generative-AI-Guidelines-20230214.pdf
约克大学	关于学生使用人工智能和翻译工具的指导意见	https://www.york.ac.uk/students/studying/assessment-and-examination/ai/
纽约大学	利用生成式人工智能进行教学	https://www.nyu.edu/faculty/teaching-and-learning-resources/teaching-with-generative-tools
芝加哥大学	打击学术不端行为第六部分：ChatGPT、人工智能与学术诚信	https://academictech.uchicago.edu/2023/01/23/combating-academic-dishonesty-part-6-chatgpt-ai-and-academic-integrity/
加利福尼亚理工学院	人工智能时代的教学资源	https://ctlo.caltech.edu/universityteaching/resources/resources-for-teaching-in-the-age-of-ai
西北大学	生成式人工智能在课程中的应用	https://ai.northwestern.edu/education/use-of-generative-artificial-intelligence-in-courses.html
得克萨斯大学奥斯汀分校	关于 ChatGPT 需要了解的五件事	https://ctl.utexas.edu/teaching-technology/5-things-know-about-chatgpt

续表

高　校	名　称	网　址
威斯康星大学麦迪逊分校	在课堂上使用人工智能的注意事项	https://idc.ls.wisc.edu/ls-design-for-learning-series/considerations-ai-classroom/
伦敦大学学院	在教学和评估中运用人工智能	https://www.ucl.ac.uk/students/exams-and-assessments/assessment-success-guide/engaging-ai-your-education-and-assessment
伦敦大学学院	在教学与学习中使用生成式人工智能（GenAI）	https://www.ucl.ac.uk/teaching-learning/publications/2023/sep/using-generative-ai-genai-learning-and-teaching
伦敦政治经济学院	伦敦政治经济学院（LSE）为教师提供的关于人工智能、评估及学术诚信的短期指导意见，以助力2022—2023学年评估期的准备工作	https://info.lse.ac.uk/staff/divisions/Eden-Centre/Assets-EC/Documents/AI-web-expansion-Feb-23/Updated-Guidance-for-staff-on-AI-A-AI-March-15-2023.Final.pdf
南洋理工大学	南洋理工大学关于在研究中使用生成式人工智能的立场	https://www.ntu.edu.sg/research/resources/use-of-gai-in-research
东京大学	关于在课堂上使用人工智能工具的政策	https://utelecon.adm.u-tokyo.ac.jp/en/docs/ai-tools-in-classes
名古屋大学	关于生成式人工智能的使用	https://en.nagoya-u.ac.jp/academics/ai/index.html
香港大学	生成式人工智能	https://tl.hku.hk/2024/08/35063/

高　校	名　称	网　址
香港科技大学	终端用户的人工智能素养：明智地运用人工智能	https://libguides.hkust.edu.hk/ai-literacy
香港中文大学	人工智能工具在教学、学习和评估中的应用（学生指南）	https://www.aqs.cuhk.edu.hk/documents/A-guide-for-students_use-of-AI-tools.pdf
香港城市大学	生成式人工智能指南	https://www.cityu.edu.hk/GenAI/guidelines.htm
香港教育大学	人工智能生成工具：指导原则	https://libguides.eduhk.hk/ai-generative-tools/guidelines
台湾大学	生成式人工智能工具用于教学的使用指南	https://www.dlc.ntu.edu.tw/en/ai-tools-en/
罗素大学集团	罗素集团关于在教育中使用生成式人工智能工具的原则	https://russellgroup.ac.uk/news/new-principles-on-use-of-ai-in-education/
爱丁堡大学	关于人工智能使用的面向学生和教职员工的指导意见	https://www.ed.ac.uk/ai/guidance
爱丁堡大学	面向学生的生成式人工智能使用指南	https://www.ed.ac.uk/bayes/ai-guidance-for-staff-and-students/ai-guidance-for-students
曼彻斯特大学	人工智能（AI）教学指导	https://documents.manchester.ac.uk/display.aspx?DocID=70286

续表

高　校	名　称	网　址
巴黎政治学院	ChatGPT：巴黎政治学院制定了相关规则，并就高等教育领域的人工智能问题展开了思考	https://www.sciencespo.fr/fr/actualites/sciences-po-fixe-des-regles-claires-sur-lutilisation-de-chat-gpt-par-les-etudiants/
墨尔本大学	评估与人工智能	https://melbourne-cshe.unimelb.edu.au/ai-aai/home/ai-assessment
悉尼大学	人工智能在教育领域的应用	https://canvas.sydney.edu.au/courses/51655
昆士兰大学	人工智能在作业中的使用	https://guides.library.uq.edu.au/referencing/ai-tools-assignments
麦吉尔大学	在教与学中使用生成式人工智能	https://deptkb.mcgill.ca/display/TLK/Using+Generative+AI+in+Teaching+and+Learning
鲁汶大学	负责任地使用生成式人工智能	https://www.kuleuven.be/english/genai
奥克兰大学	在课程作业中使用生成式人工智能的教师指南	https://teachwell.auckland.ac.nz/resources/generative-ai/ai-tools-in-coursework/
利物浦大学	生成式人工智能在教学、学习与评估中的应用	https://www.liverpool.ac.uk/centre-for-innovation-in-education/digital-education/generative-artificial-intelligence/

续表

高 校	名 称	网 址
悉尼科技大学	为学科制定关于生成式人工智能的指导方针	https://educationexpress.uts.edu.au/collections/artificial-intelligence-in-learning-and-teaching/resources/developing-guidelines-for-generative-ai-in-your-subject/
北京师范大学、华东师范大学	生成式人工智能学生使用指南	https://www.ecnu.edu.cn/info/1094/67178.htm
上海交通大学	规范学生使用人工智能工具的教师指南	https://ctldnew.sjtu.edu.cn/news/detail/1143
上海科技大学	生成式人工智能使用指南	https://ai.shanghaitech.edu.cn/2024/0327/c14346a1093334/page.htm
中国传媒大学文化产业管理学院	大学生正确使用生成式人工智能倡议	https://scim.cuc.edu.cn/_t482/2023/1025/c8345a213131/page.htm
中国传媒大学继续教育学院	继续教育学院关于学历继续教育本科毕业论文（设计）使用人工智能的规定	http://mdedu.cuc.edu.cn/mdedu2021/2021content.aspx?id=15190
湖北大学	关于开展我校 2024 届普通本科毕业学生毕业论文（设计）检测工作的通知	https://jwc.hubu.edu.cn/info/1061/7877.htm

续表

高　校	名　称	网　址
山西农业大学	关于对 2024 届本科毕业设计（论文）试行 AIGC 检测的通知	https://jwc.sxau.edu.cn/info/1006/8821.htm
天津科技大学	关于 2024 年本科生毕业设计（论文）查重和 AIGC 检测的通知	https://jw.tust.edu.cn/ggtz/65b256ce743a41fa82c3533e16009807.html
福州大学	关于对 2024 届本科毕业设计（论文）试行 AIGC 检测的通知	https://jwch.fzu.edu.cn/info/1039/13415.htm

附录 3:《高等教育中的ChatGPT 和人工智能: 快速入门指南 》中的使用建议①

一、谨慎而有创造性地使用ChatGPT

（1）创造机会让教职员工、学生和其他利益相关者讨论ChatGPT
　　　对高等教育机构的影响，并共同制定适应AI的战略；

（2）为学生和教师提供关于如何使用ChatGPT，以及何时使用或
　　　何时不能使用ChatGPT的明确指导；

（3）将ChatGPT的使用与课程学习效果相结合；

（4）审查现有的评估及评价形式，以确保其适用性；

（5）审查并更新与ChatGPT及其他AI工具相关的学术诚信政策；

（6）对教师、研究人员及学生进行培训，以提高他们向ChatGPT
　　　提问的水平。研究人员发现，越精心地设计提问，ChatGPT
　　　回答效果越好。

二、构建理解和使用ChatGPT 的能力

（1）聚焦ChatGPT和AI的新课程，提升高等教育机构的AI研发
　　　能力，并为学生提供前沿知识；

（2）更新现有课程，将AI素养、AI伦理、AI能力纳入教学
　　　内容；

---

① UNESCO, "ChatGPT and Artificial Intelligence in Higher Education: Quick Start Guide,"
2023, https://unesdoc.unesco.org/ark:/48223/pf0000385146.

（3）培训教师，确保他们在教学中发挥ChatGPT和其他AI工具
的优势，而不仅仅是复制聊天机器人或AI工具所提供的
内容；

（4）可以在学院内部、本机构或上级机构中为教师提供同行支
持和辅导，如分享在教学和研究中使用ChatGPT的实践经
验，提高教师的技能水平。

三、进行人工智能审核

（1）了解AI的使用现状。了解什么是数据驱动的AI，需要收
集哪些数据，以及如何进行数据处理；目前高等教育机构
中使用了哪些类型的AI，它们如何支持高等教育机构的运
作；在制度层面，存在哪些相关政策或法规，高等教育机
构需要考虑哪些外部政策或法规。

（2）决定使用哪种AI技术。明确AI技术的服务领域，如用于学
生服务、评估或研究，以及AI技术增值；高等教育机构对
AI工具开源、可访问性、商业化的立场是什么；高等教育
机构如何确保所有人都能访问和使用AI工具。

（3）监控AI有效性和公平性。明确AI在满足既定需求方面的有
效性，以及对有效性的衡量标准；了解所收集的数据可否
供高等教育机构使用，包括如何使用以及多长时间收集一
次数据；确定AI在多大程度上克服或解决了公平问题。

English Version

1

Transformations Faced by University Educators in the Age of Intelligence

Challenges and Opportunities of AI for Higher Education Educators

AI is evolving into a general-purpose technology that is changing the direction of global higher education institution (HEI) development and the teaching and research environment of university educators at an unprecedented speed.

On the one hand, the requirements for jobs in the industry are undergoing tremendous changes. According to "The Future of Jobs Report 2023" published by the World Economic Forum, AI is projected to lead to the displacement of 83 million jobs globally by 2027, while simultaneously creating 69 million new jobs. Jobs in the fields of AI and machine learning have become the most rapidly growing jobs from 2023 to 2027, AI and big data has been identified as one of the most important skills to be developed in the future.[1] The Ministry of Human

---

[1] World Economic Forum, "The Future of Jobs Report 2023," 2023, https://www.weforum.org/publications/the-future-of-jobs-report-2023/.

Resources and Social Security of the People's Republic of China, in conjunction with relevant departments, has released five batches of 74 new occupations since 2019, and it was found that these new occupations are densely distributed in smart information technology-related industries. It can be seen that China's future employment market will also further increase the demand for talents related to the application of intelligent technology industries.

In view of the universality of the application of AI industry, the United Nations Educational, Scientific and Cultural Organization (UNESCO) in 2024 released the AI competency frameworks for both students and teachers, which listed AI literacy as the necessary literacy for students and teachers, pointing out that teachers and students need to master the knowledge, skills and attitudes related to AI, so that in the field of education and other areas, they can understand and use AI in a safe and meaningful way.[①] In order to adapt to the skill changes in the demand of employers in the age of intelligence, HEIs need to adjust the objectives and training content of the cultivation of talents in a timely manner.

On the other hand, AI is triggering a huge change in the methods of knowledge acquisition, application, and innovation of educators and learners in HEIs, which highlights the needs of the time for innovation

---

① UNESCO, "AI Competency Framework for Teachers," 2024, https://unesdoc.unesco.org/ark:/48223/pf0000391104; UNESCO, "AI Competency Framework for Students," 2024, https://unesdoc.unesco.org/ark:/48223/pf0000391105.

in teaching and scientific research of university educators. At the 2024 Global MOOC and Online Education Conference (GMC), the "Infinite Possibilities: Digital Development Report of Global Higher Education (2024)" and the "Digital Development Index of Global Higher Education (2024)" were released. It was pointed out that digital technology is triggering a comprehensive transformation of HEIs throughout the world, and that the global higher education has entered the "Year of Smart Education."

1. AI Triggers Changes in Knowledge Acquisition by Educators and Learners in HEIs

Knowledge is the crystallization of human wisdom. Traditionally, university educators and learners acquire knowledge primarily through oral instruction, educator–learner interaction, learner–learner collaboration, and engagement with various traditional media (such as books, newspapers, magazines, electronic audiovisual products, and the Internet). Knowledge acquisition typically takes place within specific time and space constraints, with the amount of accessible knowledge limited by the medium used. Although the Internet has, to some extent, overcome traditional limitations of time, space, and knowledge volume, it is still influenced by the quality of platforms or search engines, as well as individual search skills. The advent of AI technology has significantly enriched methods available for knowledge acquisition in HEIs, enhancing the efficiency of this process. With intelligent search and recommendation systems, AI can precisely target knowledge bases according to the specific needs of educators and learners in teaching and learning.

Generative AI (GenAI) leverages conversational AI technology to extract necessary knowledge from vast databases of digital books, online articles, and other media sources. It achieves this by undergoing extensive text-based training, combining both supervised and unsupervised learning techniques, and enabling smooth interactions with users. This greatly facilitate the way in which humans acquire knowledge. For university educators, this convenient access to information can significantly boost research efficiency and support the creation of diverse educational resources, including lecture plan design, case study writing, and exercise or homework generation. Tools such as virtual teaching assistants, intelligent learning companions, chatbots, and automated assessment systems can assist educators in tasks such as designing lecture plans, refining course materials, preparing quizzes and exams, as well as automatic grading. For college students, GenAI facilitates dialogue-based learning, quick searches, rapid Q&A, personalized learning experiences, and real-time assessments, thereby enhancing the efficiency and quality of both teaching and learning. The integration of AI in higher education has facilitated the emergence of "educator–AI–learner" teaching dynamic, where educators, learners, and AI collaborate in the educational process.

2. AI Triggers Changes in Knowledge Application by Educators and Learners in HEIs

AI promote dynamic, intelligent, and personalized development in the application of knowledge by educators and learners in HEIs, thus promoting the innovative development of teaching and research in HEIs. In the traditional education model, the application of knowledge is mainly reflected in educators' lectures and learners' homework, and the teaching content is often fixed and singular. With the help of AI, educators can accurately analyze the learning progress and knowledge mastery of learners based on their learning data and adjust the teaching content and methods in real-time to provide learners with personalized learning experiences. This dynamic mode of knowledge application helps improve teaching efficiency and promotes independent learning and deep learning among learners. As the main field of knowledge application in HEIs, research will also be strongly affected by the application of AI technology. In traditional scientific research, researchers often spend substantial time on data collection, organization and analysis, with their work frequently constrained by time and resources. AI, however, can efficiently process vast amounts of data, automatically generate analytical models, and even propose new research hypotheses. This significantly shortens the research cycle, opens up new research directions, and broadens the application of knowledge for researchers.

The application of AI in education has promoted the automation of information collection, interdisciplinary integration of knowledge, and visualization of knowledge presentation. Personalized recommendations and virtual assistants based on intelligent technology have made ubiquitous learning more popular. The application of knowledge has become more personalized, and learning and knowledge applications on the basis of real situations have become increasingly common. The "educator–learner" binary structure in HEIs is shifting to the "educator–AI–learner" ternary structure. This shift promotes the ubiquity of the learning space, meets the personalized needs of the learning process with full coverage, and creates a collaborative learning space for human and AI. Implementing a closed-loop collaborative learning mechanism, known as "human-in-the-loop," and developing a theoretical model of computation that integrates data-driven induction, knowledge-guided deduction, and feedback-based epiphany will be essential for the future "educator–AI–learner" integrated learning space.

The new generation of AI technology has the characteristics of deep learning, cross-border integration, human–AI collaboration, openness to collective intelligence, and autonomous control. However, its explainability, system deviation, data security, data privacy, and other issues have brought unprecedented problems to industry and society. If used inappropriately, AI educational applications can lead to numerous negative impacts, such as the marginalization of educators' roles, the isolation of student learning, fragmentation of knowledge systems, risks of privacy breaches, discrimination and bias, ethical concerns, imbalances in academic integrity and fairness, the alienation of educational relationships, security challenges, knowledge blind spots and information silos, a lack of content accuracy, weakening of students' higher-order thinking skills, and a growing digital divide.

3. AI Triggers Changes in Knowledge Innovation by Educators and Learners in HEIs

Knowledge innovation is a critical foundation for the continuous advancement of human civilization. It involves the discovery of new theories or application domains, the scientific explanation of known phenomena, applied research on existing theories, or the integration, refinement and development of established theoretical and application systems. Traditionally, knowledge innovation in HEIs is driven by researchers using conventional media tools, and achieved through years of study, research, knowledge construction and academic exchange. The advent of AI introduces a new catalyst for knowledge innovation, enabling researchers to generate hypotheses, design experiments, compute results, and explain mechanisms more efficiently. AI is particularly useful in assisting researchers with repetitive validations and trial-and-error processes under various assumptions, thereby boosting research productivity, streamlining the innovation process, and lowering the barriers to innovation. This frees researchers in HEIs from tedious tasks of data analysis and model building, saving more time and energy for idea generation. The human–AI collaborative model can accelerate scientific breakthroughs and open up research areas that were previously inaccessible.

Large language models (LLMs) interact with humans via natural language and integrate various applications through plug-ins, serving as a gateway connecting the triad of Cyber–Physical–Human (CPH) space. Additionally, intelligent agents—AI models capable of perceiving their environment, making autonomous decisions, and taking action—are combined with GenAI base models to form the cutting-edge vertical domain of AI agents. These AI agents, leveraging content synthesis, exhibit autonomous behaviors such as information retrieval, human–AI dialogue, task execution, and logical reasoning. When applied to scientific research, AI agents can enhance human learning capabilities and expand research paradigms in unexplored domains.

Through advanced algorithms, particularly neural network architectures like Transformer, AI learns the symbiotic relationships between words from vast corpora, enabling natural language synthesis. GenAI systems, such as ChatGPT—built with Transformer models at their core—enhance their nonlinear mapping capabilities by understanding word and sentence correlations in large datasets. By scaling up model sizes and surpassing the so-called "Feynman Limit" according to the Scaling Law, these systems quickly acquire the ability to generate language with contextual relevance, forming coherent and meaningful sentences that can generate valuable content and knowledge.

With its powerful natural language processing and generative capabilities, GenAI can create personalized content tailored to users' needs and preferences, as well as how to comprehend and process vast corpora. It means that knowledge production is no longer limited to individual human capabilities and time investment; instead, it can be achieved efficiently and at scale through algorithms. This shift not only accelerates the efficiency and speed of knowledge production but also opens possibilities for the integration, dissemination, and innovation of human knowledge, driving higher education toward greater intelligence and informatization.

However, the quality of AI-generated knowledge is constrained by factors such as learning algorithms, the quality of training data, model size, and the design of prompts. AI systems are prone to issues like over-generation (producing content that blends real and fabricated information) or under-generation. The emergence of GenAI positions intelligent machines as auxiliaries in knowledge production, challenging individual learners' critical thinking, judgment making, and independent learning. This shift may also impact ethical and moral perspectives, raising concerns about the marginalization of human intelligence by AI. Furthermore, AI-driven knowledge production introduces challenges such as algorithmic bias, data privacy concerns and ethical risks, inaccuracies in content, and threats to academic integrity and fairness. Therefore, university educators and learners must be aware of the limitations of AI when using it for knowledge production.

By understanding its drawbacks, they can maximize its benefits while ensuring the fairness and sustainability of knowledge production.

Positioning and Skill Requirements for University Educators in the Age of Intelligence

As the core drivers of higher education development, university educators play a pivotal role in talent cultivation, scientific research and innovation, social services, and cultural heritage within HEIs. In the age of intelligence, university educators must accurately recognize changes, respond scientifically, and adapt proactively. This involves reimagining talent cultivation goals, pathways, and support systems across various disciplines, aiming to produce graduates with strong AI literacy who can contribute to a harmonious, healthy, and sustainable future society.

1. The Role and Positioning of University Educators in the Age of Intelligence

University educators typically have undergone extensive academic training, with a deep reservoir of professional knowledge and a strong potential for scientific research and innovation. Unlike primary and secondary school educators who focus on foundational teaching, university educators take on multiple roles within higher education institutions. They are not only key mentors guiding students toward personal and academic success but also pioneers exploring the frontiers of their disciplines. Additionally, they actively contribute to knowledge production, engage in societal practice, drive change, and often serve as custodians and innovators of national and local cultural heritage. The comprehensive quality of university educators significantly impacts knowledge production and dissemination, talent cultivation, and research innovation in HEIs.

Although intelligent technologies may play its unique advantages in higher education, the existence of university educators is still irreplaceable. University educators are not only responsible for knowledge inheritance and innovation, promoting science and research, but more importantly, cultivating future successors with high comprehensive quality, leading learners to use intelligent tools to improve learning methods, enhance learning efficiency, practice technological ethics, advocate lifelong learning, better serve society, and inherit human civilization.

In fostering learners' core competencies (e.g., humanistic values, social responsibility, national identity, cross-cultural communication) and 21st-century skills (e.g., creativity, critical thinking, complex problem-solving, communication, and collaboration), the role of university educators cannot be replaced by AI. However, to remain competent in the smart era, university educators must continuously learn, update their concepts of education, and acquire new skills in teaching and research.

2. New Requirements on Teaching Skills for University Educators in the Age of Intelligence

In terms of teaching, university educators need to adapt swiftly to the human–AI collaborative teaching environment, understand the learning preferences and styles of modern students, and adjust teaching objectives, content and methods accordingly. They need to become proficient in using intelligent teaching tools and platforms, embracing innovative pedagogies that emphasize interactive and personalized learning. This involves not only stimulating learners' interest and motivation but also developing their comprehensive qualities and innovative abilities. Moreover, educators should prioritize nurturing learners' sense of social responsibility, ethical awareness, and mission. In the emerging "educator–AI–learner" classroom ecosystem, educators must rethink their roles, leveraging the strengths of AI tutors for instruction and problem-solving while harnessing their unique human capabilities in cultivating higher-order thinking, interpersonal communication and collaboration skills. In addition, educators should embrace intelligent technologies to reform teaching evaluations, using data-driven assessments to support teaching and learning goals more effectively. This approach can lead to more scientific, efficient, and meaningful evaluations, ultimately enhancing the overall educational experience.

3. New Requirements on Scientific Research and Innovation Skills for University Educators in the Age of Intelligence

In terms of scientific research and innovation, university educators need to master fundamental knowledge and skills in the field of AI. This will enable them to leverage various intelligent technologies to keep themselves and their teams updated with the latest knowledge in their professional fields. University educators with the capability and resources can also collaborate through industry–academia–research partnerships both inside and outside the HEI to develop intelligent tools or platforms required by their disciplines, such as domain-specific LLMs. This collaboration will facilitate better scientific research and innovation within disciplinary and interdisciplinary fields, allow for more scientifically informed identification of meaningful research problems, support the comprehensive collection of necessary research data and evidences, and streamline the research process in a human–AI collaborative environment. Ultimately, intelligent technologies will help present research outcomes more effectively, achieving the goal of scientific innovation.

2

Part II

The Concept and Connotation of AI Literacy of University Educators

Definition of AI Literacy of University Educators

In Chinese, the word "literacy" (素养) originated from Liu Biao's Biography in Book 2 of Volume 74 of *Hou Han Shu* (*Book of the Later Han*). "Those who have higher levels of literacy can be motivated and bring others together to participate in or support an action or decision if benefits or advantages are shown to them." The term refers to daily personal accomplishment (修养), which in a broader sense encompasses all aspects of moral character, appearance, knowledge and ability. In English, the term is usually interchangeable with "competence."

In 1997, the Organization for Economic Co-operation and Development (OECD) launched the "Definition and Selection of Competencies" (abbreviated as DeSeCo) project, which focused on "literacy" for nine years and finally defined literacy as "the ability to meet complex demands, by drawing on and mobilizing psychosocial

resources (including skills and attitudes) in a particular context."①
Meanwhile, it is pointed out that literacy is contemporary, holistic,
developmental, and measurable, i.e., literacy is dependent on specific
contexts, and all abilities and skills that help individuals adapt to society
or solve complex problems can be called literacy (contemporary);
literacy is generally the integration of knowledge, abilities, and attitudes
(holistic). Literacy can be achieved through specific educational means
(developmental), and can be measured through comprehensible,
operational, and assessable indicators (measurable).

1. Concept and Connotation of AI Literacy of University Educators

The Teachers Law of the People's Republic of China sets out
thebasic elements that should be present in teachers' literacy. Teachers
should, first of all, abide by laws and professional ethics and become a
mentor and friend to students. Teachers should also pay close attention
to and take care of each student while completing their teaching tasks,
respect students' dignity, and promote their all-round development.
In addition, teachers should improve their own ideological level and
teaching competence to enhance their professional skills and literacy
so as to better fulfill the mission of educating people. The connotation
of teachers' literacy is also enriched by the continuous development of

---

① OECD, "The Definition and Selection of Key Competencies: Executive Summary," 2005, https://www.deseco.ch/bfs/deseco/en/index/02.parsys.43469.downloadList.2296.DownloadFile.tmp/2005.dskcexecutivesummary.en.pdf.

society.[1] The requirements for teachers' literacy in the new era cover varied aspects, including ideological and political literacy, cultural literacy, professional literacy, and modern literacy. Ideological and political literacy is one of the essentials, which requires teachers to have correct outlooks on history, politics, values and life, etc. Cultural literacy requires teachers to have broad knowledge and cultural heritage to be able to appreciate the charm of culture and to impart knowledge and culture to students. Professional literacy requires educators to have solid subject knowledge and knowledge of education theories and to be able to flexibly use various teaching resources and means to effectively promote students' learning. Modern literacy requires teachers to be able to master modern educational technology and be adept at innovation and application to achieve quality education.[2]

2. The Proposal of the Concept of AI Literacy

As humanity enters the age of intelligence, AI literacy is gradually becoming one of the most important literacies for individual survival and development. The concept of AI literacy was first proposed in the 1970s, when the focus was on the literacy composition of AI professionals and technicians.[3] Since 2017, with the increasing

[1] 中华人民共和国教育部：《中华人民共和国教师法》，2009 年，http://www.moe.gov.cn/jyb_sjzl/sjzl_zcfg/zcfg_jyfl/ tnull_1314.html。
[2] 许扬：《新时代高校思想政治理论课教师素养提升研究》，广西师范大学硕士学位论文，2023 年。
[3] Agre, P. E., "What to Read: A Biased Guide to AI Literacy for the Beginner," MIT Artificial Intelligence Laboratory Working Papers, WP-239, 1972.

impact of AI on human society, the topic of improving the AI literacy of modern citizens has increasingly attracted attention from all walks of life. Scholars at home and abroad have defined the concept of AI literacy. For example, Long and Magerko noted that AI literacy originates and extends from technology-related literacy (information literacy, digital literacy, computational literacy, data literacy, etc.), which includes "a set of competencies that enable individuals to critically evaluate AI technologies; communicate and collaborate effectively with AI; and use AI online, at home, and in the workplace."[①] According to Wong et al., AI literacy consists of "AI concepts, AI applications, and AI ethics/safety," with "AI concepts" referring to basic AI knowledge and their origins, "AI applications" referring to AI technologies, and "AI ethics/safety" referring to the ethical challenges and safety issues in the application of AI. [②]

In December 2021, the UNESCO referred to the importance of AI literacy at a conference on the theme of "AI in Education" and defined AI literacy broadly as the knowledge, understanding, skills, and values about AI. It was argued that all citizens need to be AI literate—this has become the basic grammar of the 21st century. In 2024, the UNESCO published the "AI Competency Framework for

---

① Long, D., Magerko, B., "What Is AI Literacy? Competencies and Design Considerations," Proceedings of the 2020 CHI Conference on Human Factors in Computing Systems, 2020, pp. 1-16.
② Wong, G. K. W., Ma, X., Dillenbourg, P. et al., "Broadening Artificial Intelligence Education in K-12: Where to Start?," ACM Inroads, 2020, 11(1): 20-29.

Teachers," which states that teachers in the age of intelligence should be equipped with AI competencies in five aspects, namely, human-centered mindset, ethics of AI, AI foundations and applications, AI pedagogy, and AI for professional development.[①] The framework highlights the direction for global educators of all levels and types to improve their AI literacy.

3. Definition of AI Literacy of University Educators

University educators in the age of intelligence are encountering new opportunities and challenges in teaching, research and innovation, social services, cultural heritage, etc. Among these areas, in the key fields of teaching and research, AI intervention is fundamentally reshaping and innovating the objectives, content, methods, approaches, and evaluation systems. University educators must confront these significant changes by updating their educational concepts, mastering and updating continuously their AI knowledge and skills, leveraging AI to enhance their innovative capabilities, and upholding sound values and ethical principles in relation to AI.

Given the significance and specificity of the university educators, exploring the concept, connotation, and improvement of university educators' AI literacy is crucial for the transformation of HEIs in the age of intelligence. We investigated the concept and content framework

---

① UNESCO, "AI Competency Framework for Teachers," 2024, https://unesdoc.unesco.org/ark:/48223/pf0000391104.

of AI literacy/competence worldwide and took into account the work characteristics of university educators, based on which, we define AI literacy of university educators as the competencies they should possess in the age of intelligence. These include a commitment to the advanced concepts of education (what it means to be an educator), mastery of AI knowledge (what contributes to an educator), transformation of teaching and research models (how to evolve as an educator), and the ability to assume social responsibility (what it means to be a responsible educator). Specifically, AI literacy of university educators encompasses five key dimensions: the advanced concepts in the age of intelligence, basic knowledge of intelligent education, human–AI collaborative teaching capability, scientific research and innovation empowered by AI, and the human-centered values of technology for greater good. Among these, the advanced concepts serve as the guide, knowledge as the foundation, capability as the core, innovation as the focus, and values as the essence. These five dimensions complement and integrate with each other (see Figure 1).

A Commitment to the Advanced Concepts of Education

What it means to be an educator

Mastery of AI Knowledge

What contrbutes to an educator

Transformation of Teaching and Research Models

How to evolve as an educator

The Ability to Assume Social Responsibility

What it means to be a responsible educator

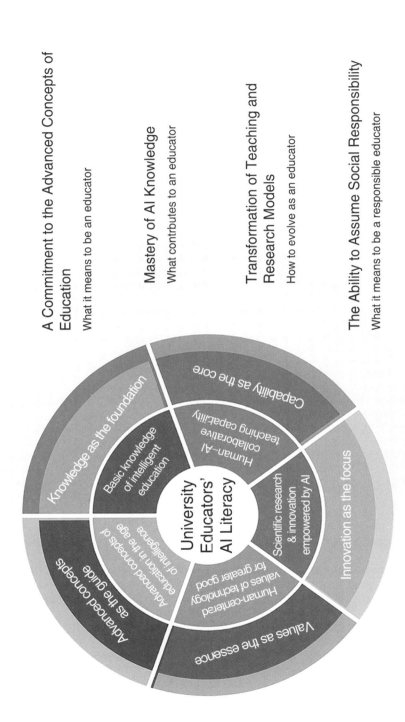

Figure 1 Concept of University Educators' AI Literacy

Connotation of AI Literacy of University Educators

The connotation of AI literacy of university educators includes the following points.

1. Advanced Concepts of Education in the Age of Intelligence

In the age of intelligence, the model of human knowledge production and dissemination is undergoing significant changes. Human–AI collaboration/interaction has become the norm for both university educators and learners in learning and work. The easy access to various educational resources and intelligent recommendations are fundamentally changing the model of education and the educator–learner relationship in HEIs. As a result, diverse and mixed-quality information has more opportunities to influence the worldview, values, and perspectives of both university educators and learners. Therefore, university educators need to update their educational concepts in a timely manner. Besides requiring oneself to become a qualified educator in the new era with the standards of "morality and integrity," university educators should also integrate words and deeds into the entire process of education, pay more attention to the character cultivation and personality shaping of learners in the age of intelligence, take "developing the whole person" as the talent cultivation goal, empower the education process through technology, and reconstruct the education scene and behavior. University educators need to realize that in the age of intelligence, fostering virtue through education is the foundation, innovative thinking is the goal, personalized education is the key, combination of theory with practice is the path, and technological empowerment is the support. Table 1 presents the content description of the "advanced concepts of education in the age of intelligence" dimension.

Table 1 Content Description of "Advanced Concepts of Education in the Age of Intelligence" Dimension

Primary dimension	Secondary dimension	Description
Advanced concepts of education in the age of intelligence	Foundation: To foster virtue through education	Understanding the significance of cultivating talents with virtue in HEIs in the age of intelligence
	Goal: Innovative thinking	Clarifying that the development of students' innovative thinking is an important goal in the talent cultivation in HEIs in the age of intelligence
	Key: Personalized education	Understanding that personalized education and teaching students in line with their abilities are the keys to reform the talent cultivation in HEIs in the age of intelligence
	Path: Combination of theory and practice	Advocating and practicing the close combination of theory and practice in the process of talent cultivation in the age of intelligence
	Support: Technological empowerment	Recognizing and embracing the positive role of technology in teaching and scientific research in HEIs in the age of intelligence

The advanced concepts of education in the age of intelligence refer to university educators' deep understanding of "cultivating socialist builders and successors with comprehensive development in morality, intelligence, physical fitness, aesthetics, and labor skills" as the fundamental task of Chinese HEIs. They need to be aware of the new challenges faced by moral education in HEIs in the age of intelligence and the importance of cultivating talents with virtue for individuals, families, society and nation. University educators should make it clear that intelligent technology can empower human production, life and learning, but innovative thinking will be the core competence of talents in the age of intelligence. Therefore, developing learners' innovative thinking will become an important goal of higher education. University educators should know that intelligent technology makes personalized education and tailored teaching possible, and they are the keys to the reform of talent cultivation in HEIs. University educators should understand the importance of combining theory with practice for the cultivation of innovative talents and the discovery of new knowledge in HEIs in the age of intelligence, and advocate and practice the close integration of theory and practice in teaching and research. University educators should understand the unique advantages of the intelligent technology in teaching and research, and recognize and embrace its positive role.

2. Basic Knowledge of Intelligent Education

Intelligent education aims to leverage intelligent technology to optimize the teaching and learning process in the education of different disciplines, promote the reform of teaching methods, and transform the model of talent cultivation. It strives to establish a learner-centered educational environment, providing accurate content delivery services, and achieving the goal of customizing daily and lifelong education. The core of intelligent education lies in improving teaching methods, management, and assessment through intelligent means, adapting to the diverse learning needs and ability levels of learners. Higher education educators need to master the basic knowledge of intelligent education, which includes principles of intelligent technologies, theoretical foundation of intelligent education, tools/platforms of intelligent teaching, methods of intelligent teaching and evaluation, and knowledge of intelligent education management. Table 2 presents the content description of "basic knowledge of intelligent education" dimension.

Table 2　Content Description of "Basic Knowledge of Intelligent Education" Dimension

Primary dimension	Secondary dimension	Description
Basic knowledge of intelligent education	Principles of intelligent technologies	Understanding the basic knowledge of AI and the principles of AI technologies in the disciplines

Continued

Primary dimension	Secondary dimension	Description
Basic knowledge of intelligent education	Theoretical foundation of intelligent education	Getting familiar with the general situation and trend of the application of intelligent technologies in education at home and abroad and relevant theoretical basis
	Tools/platforms of intelligent teaching	Getting familiar with intelligent tools/ platforms used in the disciplines and cases of teaching applications
	Methods of intelligent teaching and evaluation	Mastering typical methods of teaching and evaluation supported by intelligent technologies
	Knowledge of intelligent education management	Mastering the basic knowledge of using intelligent technologies to carry out educational management

University educators need to have a basic understanding of the fundamental principles of intelligent technologies, particularly the foundational knowledge and technical principles of AI related to disciplines. This includes basic knowledge related to data, algorithms, and computing power; basic knowledge about image recognition, speech recognition, and synthesis; and technical principles related to LLMs and intelligent agents. University educators should also be familiar with theories that support the application of intelligent technologies in teaching and research within their disciplines, such as the Technological Pedagogical Content Knowledge (TPACK) framework, cognitive theory of multimedia learning, generative learning theory and connectivism learning theory. They should be able to analyze successful cases of intelligent education both domestically and internationally based on these theories. Furthermore, university educators should be acquainted with representative intelligent teaching tools and platforms in their disciplines and understand classic cases of their use in HEIs worldwide, accumulating resources for intelligent teaching in their field. When conducting intelligent teaching activities, university educators should be able to select appropriate intelligent technologies based on learner characteristics and course content, improving their teaching efficiency and reducing the burden of evaluation tasks. University educators should also master the principles and standards for the use of intelligent technology in commonly used educational management systems (such as academic management systems and research management systems) at their institutions.

3. Human–AI Collaborative Teaching Capability

In intelligent education, university educators adopt human–AI collaborative teaching activities in order to practice the "learner-centered" concept in the field of discipline teaching. Through the use of AI technologies, they can better understand learners' initial level and learning needs, design reasonable learning objectives, select appropriate teaching methods, design intelligent teaching environments and learning experiences, and use diverse evaluation methods to assess the learners' performance. University educators should use various evaluation methods to carry out value-added evaluation activities. To better carry out human–AI collaborative teaching, university educators need to enhance their design thinking skills, focusing on learner-centered approaches, with an emphasis on a better understanding of learners' needs and teaching problems. Through an iterative design process, they generate creative and practically effective solutions. Design thinking particularly focuses on elements such as empathy, problem defining, designing, prototyping, testing, and iteration. Table 3 presents the content description of the "human–AI collaborative teaching capability" dimension.

Table 3　Content Description of the "Human–AI Collaborative Teaching Capability" Dimension

Primary dimension	Secondary dimension	Description
Human–AI collaborative teaching capability	Determining reasonable learning objectives	Ability to use certain intelligent technologies to better understand learners' initial levels and learning needs, and to set reasonable, multidimensional learning objectives
	Selecting appropriate teaching methods	Ability to use certain intelligent technologies to better understand characteristics of learners and contents, and to select appropriate teaching methods

Continued

Primary dimension	Secondary dimension	Description
Human–AI collaborative teaching capability	Designing intelligent teaching environments	Ability to use certain intelligent technologies to design intelligent teaching environments that are conducive to achieving learning objectives and implementing teaching methods
	Designing learners' learning experiences	Ability to use certain intelligent technologies to design learners' learning experiences that contribute to achieving learning objectives
	Designing diverse learning evaluations	Ability to use certain intelligent technologies to design diverse learning evaluations to assess whether learning objectives have been met

University educators may use intelligent technologies to optimize classroom teaching strategies. Utilizing intelligent technologies to dynamically understand learners' initial level of knowledge and skills, and learning needs before teaching could help university educators set reasonable, multidimensional learning objectives and promote the growth of learners' wisdom and abilities. University educators may rely on intelligent technologies to accurately understand learners and learning content characteristics, identify learners' blind spots and preferences, and select appropriate teaching methods. They may use intelligent technologies and interactive, perceptive devices to create virtual, real, and diverse teaching environments, designing intelligent teaching settings that support the achievement of learning objectives and the implementation of teaching methods. University educators may also use intelligent technologies to enhance learners' task execution abilities during the teaching and learning process, designing contextualized learning perceptions and immersive embodied experiences to provide activities that foster better knowledge, understanding, and inquiry. By fully leveraging the advantages of human–AI collaboration, university educators could combine intelligent and teacher-driven evaluations, designing diversified learning assessments that support comprehensive and value-added evaluations of learners. These evaluations provide dynamic, accurate, and scientific feedback, thus enabling personalized teaching.

4. Scientific Research and Innovation Empowered by AI

The combination of AI with scientific challenges and engineering difficulties, as well as the architecture of AI algorithms on top of human expertise in different disciplines, has promoted educational innovation, scientific discovery, and engineering breakthroughs through human–AI collaboration, and further promoted the exploration and implementation of intelligent systems. The paradigm of AI for Research (AI4R) will profoundly influence the future of technological development in China and the research innovation practices of university educators.

Unlike traditional ones, scientific research and innovation in the intelligent era are characterized by uncertainty, complexity, and interdisciplinary nature. Therefore, university educators need to develop their innovation capabilities in intelligent research environments, including adopting AI-based innovative actions in such key steps of scientific research as optimizing the knowledge structure of scientific research, establishing clear research directions, introducing and applying the latest research methods, executing research processes, and describing research findings. Additionally, if possible, university educators should engage in AI-related research and development in their specific disciplinary fields and promote the iterative improvement of AI systems. Table 4 presents the content description of the "scientific research and innovation empowered by AI" dimension.

Table 4   Content Description of the "Scientific Research and Innovation Empowered by AI" Dimension

Primary dimension	Secondary dimension	Description
Scientific research and innovation empowered by AI	Innovative acquisition of research knowledge	Ability to use digital intelligence technologies to innovate methods or approaches for acquiring research knowledge
	Innovative selection of research problems	Ability to use digital intelligence technologies to innovate the identification and selection of research problems

Continued

Primary dimension	Secondary dimension	Description
Scientific research and innovation empowered by AI	Innovative use of research methods	Ability to use digital intelligence technologies to innovate the selection and use of research methods, including big data methods and interdisciplinary approaches
	Innovative execution of research processes	Ability to use digital intelligence technologies to innovate the execution of research processes, with human–AI collaboration as a core feature
	Innovative presentation of research results	Ability to use digital intelligence technologies to innovate the presentation of research results

University educators can use digital intelligence technologies to quickly filter and analyze large volumes of literature data, discovering hidden research trends and correlations, thereby innovating ways or methods to acquire research knowledge. They can use digital intelligence technologies to model their own research experiences and personal interests, identify unresolved or pending problems, and uncover potential research directions. By simulating and predicting experimental results with digital intelligence technologies, educators can reduce costs and time spent on experiments, develop new experimental designs and analysis methods, and explore the selection and use of interdisciplinary approaches to help innovate research methods. Human–AI collaboration helps educators optimize the research process. Through human–AI collaboration, human and AI cooperate with each other in specific tasks to achieve higher work efficiency and creativity. The unique reasoning advantages of humans are perfectly integrated with the efficient search capabilities of AI, unlocking enormous potential for solving complex problems. In presenting innovative research results, educators can use digital intelligence technologies to innovate how results are displayed, making research findings more intuitive and vivid, breaking through the Feynman Limit, enhancing academic communication, improving public understanding, and increasing the dissemination and impact of research.

5. Human-Centered Values of Technology for Greater Good

It is stated in *The Book of Rites* that "A teacher is one who educates through practice and exemplifies virtues."[1] University educators nurture learners not only through knowledge but also through virtue. In the process of intelligent education, university educators shoulder dual missions of teaching and scientific research. Their human-centered values and moral principles in science and technology have a profound impact on learners' learning, life, and research work. Today, human beings have entered a new ecological environment characterized by tripartite space known as CPH (Cyber–Physical–Human) space, where ethical discussions are no longer limited to interpersonal relationships or the relationships between humans and nature. Instead, they involve

---

[1] "师也者，教之以事而喻诸德也。" in Chinese.

the connections between humans and artificial objects in society, which makes AI possess both technological and social attributes. Therefore, in a society with human–AI integration, humans should adhere to the value of technology for good and human-centered principles, ensuring that human values, ethics, and laws are embedded in AI products and services, thus endowing AI a social attribute. Table 5 presents the content description of the "human-centered values of technology for greater good" dimension.

Table 5   Content Description of the "Human-Centered Values of Technology for Greater Good" Dimension

Primary dimension	Secondary dimension	Description
Human-centered values of technology for greater good	Awareness of data security and privacy protection	Ability to recognize the importance of data security and privacy protection when using intelligent technologies
	Vigilance against algorithmic bias and model hallucinations	Ability to understand how algorithmic bias and model hallucinations arise, as well as their consequences, when using intelligent technologies
	Alignment with technology for good and human-centered principles	Ability to consistently adhere to technology ethics and academic integrity, ensuring that intelligent technologies serve the purpose of benefiting humanity and align with human-centered principles

Continued

Primary dimension	Secondary dimension	Description
Human-centered values of technology for greater good	Commitment to human–AI symbiosis and universal accessibility	Ability to advocate for the harmonious coexistence of human and intelligent technologies while promoting universal access to intelligent technologies
	Pursuit of equitable knowledge accumulation and sharing	Ability to utilize intelligent technologies to facilitate the accumulation of human knowledge and promote equitable access and sharing worldwide

In the environment of deep integration of intelligent technologies and education, university educators should actively reflect on and rationally examine the advantages and disadvantages that intelligent technologies may bring, adhere to the principle of technological progress serving the improvement of human welfare, and be alert to the possible tragedy of human destruction caused by the materialization of life. In the process of human–AI collaboration, university educators should consciously pay attention to the security of educators' and learners' personal and scientific research privacy data during collection, transmission, storage, backup, and analysis. Educators should know and deeply understand the characteristics of algorithmic models and their impacts, including the manifestations and consequences of non-interpretability, bias, illusion, and sycophancy, so that they can maintain a mindset of caution and risk awareness when using such technologies, as well as guide learners to use technology rationally. During teaching and research, university educators should consistently adhere to and guide learners in following ethical principles in technology, building the foundational principles and values of technology for good and human-centered approaches. Given that AI has social attributes, university educators should ensure that the risks of using AI in scientific and technological activities are controllable and that the achievements of science and technology benefit people, realizing the concept of universal popularization of AI. University educators should actively consider how to use intelligent technology to promote the accumulation of human knowledge in their respective disciplines, aiming for shared and universal access to frontier knowledge for all humanity.

3

Part Ⅲ

Goals, Pathways, and Supports for Enhancing University Educators' AI Literacy

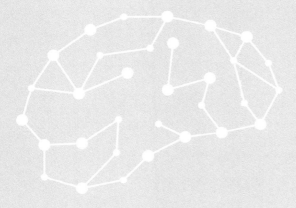

Goals for Enhancing University Educators' AI Literacy

The goals for enhancing university educators' AI literacy are to enable them, through relevant learning, communication and practical activities, to possess the advanced concepts of education in the age of intelligence, master the basic knowledge of intelligent education, develop human–AI collaborative teaching capabilities, possess AI-powered scientific research and innovation capabilities, and uphold the humanistic value of science and technology for the greater good.

Pathways for Enhancing University Educators' AI Literacy

To implement the "Opinions of the Central Committee of the Communist Party of China and the State Council on Comprehensively Deepening the Reform of Teacher Workforce Construction in the New Era" and to further integrate new technologies such as AI into teacher workforce development, the Ministry of Education of the PRC launched a pilot program in 2018 in Ningxia Hui Autonomous Region and Beijing Foreign Studies University. This program aimed to support teacher development with the integration of AI. In 2021, the scale of the pilot was expanded, covering 56 HEIs, 20 cities, and 25 districts and counties, with deeper and broader applications. The UNESCO has published several influential global reports in recent years, including "ChatGPT and Artificial Intelligence in Higher Education: Quick Start Guide" (2023), "Guidelines for Generative AI in Education and Research" (2024), and "AI Competency Framework for Teachers" (2024), which aim at guiding university educators to better integrate AI into higher education.

It is evident that focusing on enhancing AI literacy of university educators, especially addressing weaknesses, is a crucial step in driving global change in higher education in the age of intelligence. This includes promoting the use of intelligent assistants (platforms, systems, resources, tools, etc.) by educators to foster reforms in teaching and research methods, and learning approaches, thereby reducing educators' workloads while empowering them. These efforts are universally recognized as essential measures for adapting to the changes in higher education in the age of intelligence. The main pathways for enhancing university educators' AI literacy are as follows.

1. Updating the Contents and Forms of University Educator Training

As the deepening of AI applications in higher education and research, the human resources departments and those responsible for university educators' professional development should organize in-service training around the theme of "AI + Education," targeting different groups of educators (newly hired educators, middle-aged and/ or young educators with some teaching and research experience, and senior educators with many years of experience). The focus should be on enhancing university educators' practical skills in using intelligent technologies for teaching, research, and innovation. To improve the five dimensions of AI literacy among university educators, it is suggested that colleges and universities organize the following activities:

i. Through faculty meetings, teacher seminars, training for new and experienced educators, mentor programs, and other activities at the school and departmental levels, new concepts for education in the age of intelligence should be promoted. This includes sharing successful cases from representative HEIs and disciplines to continually raise awareness of modern educational philosophies among all educators.

ii. Invite experts in the field of intelligent education to deliver keynote speeches on the development and latest progress of AI, and basic knowledge and skills for using AI in education. Institutions with necessary resources could develop MOOCs or micro-course resources for intelligent education, which are accessible on the school's learning

platform for self-study by educators. The completion of these courses can be incorporated into the teacher training credit recognition system established by the human resources department or teacher development center.

iii. Teaching management departments for undergraduate and graduate programs should organize university educators to systematically study representative AI teaching application guidelines worldwide and understand the pathways, key considerations, and best global practices for implementing human–AI collaborative teaching. Institutions with necessary conditions can develop guidelines for AI educational applications tailored to their disciplinary characteristics and the needs of educators and learners, providing guidance for educators to carry out human–AI collaborative teaching. Innovative teaching workshops, education reform project proposals, and experience-sharing activities should also be organized to allow educators to directly engage in the collaborative teaching process.

iv. Research management departments, information technology centers (or relative technical support departments), and teacher training departments should gather typical cases of AI-assisted scientific research worldwide. These can be shared through various teacher training events, project application mobilization meetings, and research-focused seminars. Through these activities, educators will quickly become familiar with intelligent technologies that can assist their research and learn how to use them effectively. Additionally, workshops and site visits to enterprises and research institutions can help educators experience the process of AI-empowered scientific research innovation firsthand.

v. Through collaboration among teaching management, research management, and teacher training departments, HEIs can invite interdisciplinary experts in technology philosophy, technology ethics, and AI ethics to introduce basic knowledge and the best practices in these fields. Activities can be in form of lectures, case studies, thematic debates, and workshops to inspire educators to reflect on the concepts, connotations, and ethical practices of AI for greater good and human-centered values. These activities will guide educators to integrate these ethical perspectives into their thinking and actions in both teaching and research.

Trainings for enhancing university educators' AI literacy should be as diverse and multifaceted as possible to meet various purposes and training needs. They can include traditional lectures focused on imparting AI knowledge and teaching experiences, as well as workshops, practical activities, or short courses aimed at developing AI application skills. They can also involve heuristic, dialogue-based seminars and debates. Trainings can be face-to-face, hybrid, or fully online, depending on the context and needs of the participants.

2. Innovating the Contents and Forms of Teaching and Research Studios for University Educators

Teaching and research activities (also known as the study of teaching or jiaoyan) at HEIs serve as a vital pathway to promote the professional development of university educators and enhance their teaching capabilities. The teaching and research studios (also called jiaoyan office[①]) play a significant role in organizing these activities. Originating in the 1950s, teaching and research studios were typically organized by subject or course and served as basic teaching units in HEIs. These studios are vital grassroots organizations dedicated to the educational reform and professional development of educators. They facilitate the mentoring process[②] between experienced and novice educators and help improve overall teaching quality. Traditionally, teaching study group activities are conducted at designated time and locations, where educators gather to discuss challenges in education and teaching, share experiences and work collectively to improve educational outcomes. The primary goal of these activities is to enhance the quality of education and teaching. To better align with the ongoing transformations in higher education driven by the age of intelligence, there is a pressing need to innovate the contents and forms of teaching study activities. Regarding contents, given the profound impact of AI on the daily teaching and learning behaviors of both educators and

---

① "教研室" in Chinese.
② "传帮带" in Chinese.

learners, it is essential for HEIs and their departments to systematically and purposefully include AI as a key topic of discussion. This should cover three dimensions: enhancing teaching strategies, improving student learning, and optimizing educational management, with a focus on exploring how AI can be integrated into higher education in a safer and more effective manner.

In terms of activity forms, beyond traditional face-to-face sessions, HEIs can leverage the advantages of information technology to conduct online teaching and research activities. In 2021, the Ministry of Education of the PRC issued a notice of "The Pilot Construction of Virtual Teaching and Research Studios," outlining an action plan for this initiative. By 2022, a total of 657 virtual teaching and research studios pilots were approved, covering three main types: professional development, course (group) teaching, and educational reform. Virtual teaching and research studios[1] represent an innovative teaching organization model that transcends the limitations of time and space. These groups integrate high-quality teaching resources from both within and outside the HEIs, creating a robust teaching and research ecosystem that supports educational reform. As a new form of grassroots teaching organization built on modern information technology, virtual teaching and research studios effectively consolidate educational resources, foster knowledge integration and innovation, and build a flexible

---

[1] "虚拟教研室" in Chinese.

teaching workforce that meets the needs of interdisciplinary education. The open nature of virtual teaching and research studios breaks down traditional barriers of time, space, and discipline, promoting collaborative teaching among educators from different institutions, fostering interdisciplinary integration, and enabling resource sharing. These advantages enable virtual teaching and research studios to optimize interdisciplinary faculty allocation and knowledge structure of research team in universities, making teaching and research work more dynamic, open, and not limited by time, space, and geography. The virtual teaching and research studios make them an essential force in promoting interdisciplinary curriculum development and training versatile talent in the age of intelligence. Appendix 1 provides the list of virtual teaching and research studios related to AI, approved by the Chinese Ministry of Education, along with a case example.

3. Training University Educators' AI-Integrated Practical Ability through Projects

Domestic and international experience has shown that project-based learning is an effective way for learners to enhance their practical abilities. By hosting or participating in various educational reform or research projects on AI+X and X+AI themes, university educators may purposefully carry out professional teaching practices and explore innovative research paths that integrate AI, thereby enhancing interdisciplinary practical abilities.

The Higher Education Department of Ministry of Education of the PRC and provincial higher education authorities, as well as undergraduate and graduate education management departments within HEIs may encourage educators from various disciplines to participate in teaching reform practice by establishing specialized educational reform projects focused on X+AI. Through these educational reform projects, educators from various disciplines in HEIs can actively explore the possibilities of applying AI in teaching, such as using AI tools to better carry out teaching design or analyze the learning process and results of learners. Educators and learners may discuss the advantages and disadvantages of AI-integrated education. Through practical case analysis of subject teaching, eduators might better understand the successful application or potential challenges of AI in education, and cultivate the ability of university educators to optimize instrucitonal plans and solve difficult instructional problems in an intelligent environment. Research outcomes based on educational reform projects (papers, books, inventions, patents, etc.) are also conducive to peer exchange and diffusion of innovaticn.

In addition, the Ministry of Education and the Ministry of Science and Technology of the PRC, and provincial scientific research management departments may set up AI+X and X+AI-related research projects to encourage university educators from different disciplines to apply. These projects promote interdisciplinary and cross-departmental collaboration among non-computer science faculty and experts in

computer science, AI, and data science, as well as industry partners. By sharing knowledge and experiences, faculty from various academic backgrounds can complement each other's strengths, ultimately fostering the integration of AI into academic research and innovation across disciplines.

Supports for Enhancing University Educators' AI Literacy

Enhancing the AI literacy of university educators requires strong supports from HEIs in four key areas: organization, insitution, resources, and environment.

1. Organizational Support

First, leadership at all levels within HEIs should recognize the urgency of educational transformation in the age of intelligence and the necessity of enhancing university educators' AI literacy. On this basis, university leaders need to engage in top-level planning and establish relevant organizational structures to advance this initiative effectively. This may include setting up dedicated task forces and AI education and research centers to ensure the systematic implementation of AI literacy enhancement. The primary goal of these entities is to facilitate AI-related training, practical application, and research for university educators. By developing relevant policies, reports, or guidelines, these organizations can guide educators across the university in integrating AI into their teaching and research activities. Furthermore, various departments such as the human resources office, undergraduate school, graduate school, teacher development center, continuing education departments, and individual colleges should collaborate across departments. This cross-functional coordination is crucial for mobilizing as many resources as possible, both internally and externally, to provide strong organizational support for AI literacy improvement among different educator groups (newly hired educators, middle-aged and/or young educators with some teaching and research experience, and senior educators with many years of experience).

2. Institutional Support

HEIs can also promote university educators' AI literacy through institutional policies. Specifically, institutions can include teacher training, teaching studies, and educational reform projects—as well as teaching awards and achievements in teaching skill competitions—as part of the criteria for faculty promotion and evaluation. This policy would incentivize educators to actively engage in intelligent education and AI-powered research activities. Additionally, HEIs in China can refer to existing domestic and international guidelines (see Appendix 2) to establish their own rules for integrating AI (especially GenAI) into educational and teaching contexts, thus standardizing teacher and student behavior. For instance, in April 2023, the UNESCO released the "ChatGPT and Artificial Intelligence in Higher Education: Quick Start

Guide," providing recommendations for higher education institutions on the use of ChatGPT (see Appendix 3). Many HEIs, both domestically and internationally, have already developed relative guidelines.

3. Resource Support

HEIs can provide substantial support for enhancing university educators' AI literacy through human, material, and financial resources. Specifically, institutions can offer free or low-cost access to AI-powered teaching and research tools/platforms, particularly those involving GenAI. These resources could be externally sourced or developed in-house. Departments such as the undergraduate and graduate teaching management, and information technology center should collaborate to support educators in developing AI assistants tailored to specific disciplines or engaging in research on LLMs within specialized fields. Moreover, HEIs should allocate adequate resources for teacher training, teaching study group activities, and educational reform projects. By doing so, institutions can offer customized pathways for enhancing AI literacy, ensuring that more university educators have the opportunity to participate in, experience, demonstrate, and promote intelligent education and AI-powered research. Encouraging interdisciplinary and cross-domain practices is also crucial.

4. Environmental Support

HEIs can enhance AI literacy among university educators by creating supportive environments. On the one hand, HEIs should leverage their unique institutional and disciplinary characteristics to select appropriate AI technologies, thus building an enabling environment for intelligent education and AI-driven research. This includes physical spaces supported by digital technology (such as smart classrooms, maker spaces, AI labs, and libraries) and digital spaces (such as teaching, research, and management platforms, as well as intelligent campus system design based on digital twins). These spaces provide educators and learners with the opportunity to engage in innovative educational and research practices. On the other hand, HEIs and related enterprises should foster a cultural environment for education and research with human–AI collaboration by promoting interdepartment and industry–academia synergy. This environment encourages industry–academia collaborations, interdisciplinary exchanges, and cross-cultural interactions among faculty and students, as well as the development and application of various digital and intelligent technologies. By doing so, HEIs can reshape traditional academic ecosystems, thereby supporting transformative changes in higher education to meet the demands of the age of intelligence.

Conclusion

The emergence of GenAI, exemplified by ChatGPT, is expected to profoundly impact the future development of human society. In response, higher education institutions around the world are actively embracing transformations to seize the opportunities and address the challenges presented by the age of intelligence. As the primary base for cultivating large numbers of innovative talents for various industries, Chinese HEIs bear the honorable mission of contributing to the national goal of becoming a global leader in education, technology, and human resources. Regardless of how technological environments evolve, the ultimate goal of education is to promote human development. Therefore, integrating AI into higher education should ultimately aim to foster college students' comprehensive development.

Aligning with global trends, Chinese HEIs are actively implementing the national strategic action for educational digitization, accelerating the digital transformation, intelligent upgrading, and integrative innovation in higher education to support its high-quality development. A critical component of this transformation is the comprehensive enhancement of university educators' competencies, especially in terms of AI-related knowledge and skills. The behaviors and attitudes of university educators towards AI will determine whether intelligent education can take place, how it will be implemented, and the ultimate effectiveness and quality of such initiatives.

The primary purpose of this red book is to remind university educators to proactively "upgrade their toolkit" to address the unprecedented challenges "the Fourth Industrial Revolution" has presented. University educators must develop a profound understanding of the significant changes in this era and recognize the necessity of continuous learning. By enhancing their AI literacy across multiple dimensions, they can have a commitment to the advanced concepts of education, mastery of AI knowledge, transformation of teaching and research models, and the ability to assume social responsibility. On the one hand, this will enable them to cultivate future generations of globally competitive leaders; on the other hand, through scientific research and innovation, they can contribute to knowledge production and civilization heritage.

Appendices

Appendix 1 List of AI-Related Virtual Teaching and Research Studios in Higher Education and Case Study

The 2022 list of the first and second batches of pilot virtual teaching and research studios approved by the Ministry of Education of the PRC clearly reveals that Emerging Engineering disciplines[①] have a significant presence, accounting for approximately 40% of the total. Notably, there are 17 projects specifically related to AI, as detailed in Table A.1.

Table A.1 List of the First Two Batches of Pilot AI-related Virtual Teaching and Research Studios Approved by the Ministry of Education, PRC

Types	Virtual Teaching and Research Studios	Institutions	Leaders
Majors	Virtual Teaching and Research Studio for Music Major (AI in Music Concentration)	Central Conservatory of Music	Yu Feng

---

① "新工科" in Chinese.

Continued

Types	Virtual Teaching and Research Studios	Institutions	Leaders
Majors	Virtual Teaching and Research Studio for Urban and Rural Planning Major (Smart Cities and Smart Planning Concentration)	Tongji University	Wu Zhiqiang
	Virtual Teaching and Research Studio for AI Major (AI+X Concentration)	Zhejiang University	Wu Fei
	Virtual Teaching and Research Studio for Architectural Electrical and Intelligent Engineering Major	Anhui Jianzhu University	Fang Qiansheng
	Virtual Teaching and Research Studio for Smart Livestock Science and Engineering Major	Northwest A&F University	Yao Junhu
	Virtual Teaching and Research Studio for Vehicle Engineering Major (Intelligent Operation and Maintenance of Rail Vehicles Concentration)	Beijing Jiaotong University	Liu Zhiming
Courses & Course Groups	Virtual Teaching and Research Studio for "101 Plan" AI Introduction Course	Zhejiang University	Wu Fei

Continued

Types	Virtual Teaching and Research Studios	Institutions	Leaders
Courses & Course Groups	Virtual Teaching and Research Studio for AI Courses	Zhejiang University of Technology	Wang Wanliang
	Virtual Teaching and Research Studio for Road Engineering Intelligent Construction and Maintenance Course Group	Changsha University of Science and Technology	Yuan Jianbo
	Virtual Teaching and Research Studio for Building Intelligence Experimental Course Group	Xi'an University of Architecture and Technology	Yu Junqi
	Virtual Teaching and Research Studio for Smart+Emerging Agricultural Science Courses	Northwest A&F University	Li Shuqin
	Virtual Teaching and Research Studio for Zhihai AI Course	Harbin Engineering University	Liu Haibo
	Virtual Teaching and Research Studio for Smart Accounting Course Group	Southeast University	Chen Zhibin
	Virtual Teaching and Research Studio for Intelligent Manufacturing Course Group	Ningxia Institute of Science and Technology	Gong Yunpeng

Continued

Types	Virtual Teaching and Research Studios	Institutions	Leaders
Educational Reform	Virtual Teaching and Research Studio for Sports and Health Teaching Research in the Smart Era	East China Normal University	Wang Xiaozan
	Virtual Teaching and Research Studio for Smart Forestry Equipment Talent Cultivation Mode Research	Nanjing Forestry University	Zhou Hongping
	Virtual Teaching and Research Studio for Smart Forestry Talent Cultivation Mode Reform	Nanjing Forestry University	Cao Fuliang

As one of the first batch of pilot units of virtual teaching and research studios approved by the Ministry of Education of the PRC, Zhejiang University's Virtual Teaching and Research Studio for AI Major (AI+X Concertration) is anchored in the AI+X Teaching Center established by the College of Computer Science and Technology in 2020. This studio has developed a "three-in-one" construction model focusing on textbook development, course sharing, and platform enhancement. Development of textbooks leads the overall direction of curriculum and platform in terms of educational concepts, cultivation mode, and instructional contents. Course sharing is a think tank that revolves around the purpose of textbook development, fully leverages the wisdom of all parties, solves various problems, and provides architecture design for platform function construction. Platform enhancement is a practical tool for testing textbooks and courses, promoting the transformation from "knowledge-based" to "ability-based," and achieving the integration of knowledge and action. These components are interwoven to foster educational innovation and talent development across disciplines. The key construction strategies are as follows.

Relying on the construction of AI+X micro-majors to expand the channels for industry–academia collaborative talent cultivation. By collaborating with industry clusters, the virtual teaching study group co-constructs AI+X micro-majors, targeting the actual needs of various industry segments. This initiative offers non-computer science majors the opportunity to learn core AI theories and their practical applications. It cultivates interdisciplinary talents with a blended knowledge structure and hands-on capabilities. The studio focuses on application-driven and problem-solving education by micro-credentials. It tightly integrates industry practices with teaching, preparing students for the demands of new technologies and emerging industries.

Developing AI+X high-quality shared resources to promote regional resource sharing. Through interinstitution partnerships, the studio systematically plans and authors a series of AI+X textbooks. These resources emphasize foundational theories in interdisciplinary fields and cutting-edge applications. supplemented with corresponding instructional videos. Additionally, the studio establishes standards for resource development and sharing mechanism, encouraging dynamic content updates and nationwide accessibility. This effort ensures that sustainable and open resources are available for AI+X talent cultivation.

Enhancing the AI+X science and education innovation platform to foster an open innovative community. The studio leverages a smart science education and training platform, featuring open-source algorithms, models, and datasets. This platform provides abundant digital learning resources and tools to enhance students' practical application skills. The studio builds a domain knowledge base, computational resource pool, and competency assessment standards, creating an open interdisciplinary and integrated innovative community. This community not only sparks the curiosity of faculty and students but also facilitates the seamless integration of theoretical knowledge, course practice, industry solutions, and product implementation.

Strengthening AI+X faculty training to enhance teaching competencies. To advance faculty growth, the studio implements a comprehensive AI+X training system. Beyond cutting-edge disciplinary content, the training includes pedagogical techniques and teaching technologies, holistically enhancing educators' teaching and research abilities. The studio introduces a "problem+solution" teaching and research model, breaking traditional disciplinary boundaries to explore interdisciplinary approaches. This problem-solving focus drives collaboration among faculty from different fields, boosting their capacity to address complex teaching challenges and fostering a culture of interdisciplinary dialogue.

Appendix 2   List of Guidelines for the Use of GenAI in Representative HEIs Worldwide

Table A.2 is a list of guidelines for the use of GenAI in representative HEIs worldwide.

Table A.2   List of Guidelines for the Use of GenAI in Representative HEIs Worldwide

HEIs	Guidelines	Websites
Harvard University, Faculty of Arts and Sciences	Initial Guidelines for the Use of Generative AI Tools at Harvard	https://huit.harvard.edu/ai/guidelines
Yale University	Guidelines for the Use of Generative AI Tools	https://provost.yale.edu/news/guidelines-use-generative-ai-tools
Stanford University	Generative AI Policy Guidance	https://communitystandards.stanford.edu/generative-ai-policy-guidance
Massachusetts Institute of Technology	Teaching & Learning with ChatGPT: Opportunity or Quagmire? Part III	https://tll.mit.edu/teaching-learning-with-chatgpt-opportunity-or-quagmire-part-iii/
University of Oxford	Four Lessons from ChatGPT: Challenges and Opportunities for Educators	https://www.ctl.ox.ac.uk/article/four-lessons-from-chatgpt-challenges-and-opportunities-for-educators
University of Cambridge	How We Use Generative AI Tools	https://www.communications.cam.ac.uk/generative-ai-tool-guidelines

Continued

HEIs	Guidelines	Websites
Carnegie Mellon University	Generative AI Tools FAQ	https://www.cmu.edu/ teaching/technology/aitools/ index.html
University of California, Berkeley	Understanding AI Writing Tools and Their Uses for Teaching and Learning at UC Berkeley	https://teaching.berkeley.edu/ understanding-ai-writing-tools-and-their-uses-teaching-and-learning-uc-berkeley
University of California, Los Angeles	Generative AI for Teaching and Learning at UCLA	https://teaching.ucla.edu/ resources/ai_guidance/
University of Southern California	Instructor Guidelines for Student Use of Generative AI for Academic Work	https://academicsenate. usc.edu/wp-content/ uploads/sites/6/2023/02/ CIS-Generative-AI-Guidelines-20230214.pdf
New York University	Teaching with Generative AI	https://www.nyu.edu/faculty/ teaching-and-learning-resources/teaching-with-generative-tools
University of Chicago	Combating Academic Dishonesty, Part 6: ChatGPT, AI, and Academic Integrity	https://academictech. uchicago.edu/2023/01/23/ combating-academic-dishonesty-part-6-chatgpt-ai-and-academic-integrity/
California Institute of Technology	Resources for Teaching in the Age of AI	https://ctlo.caltech.edu/ universityteaching/resources/ resources-for-teaching-in-the-age-of-ai

Continued

HEIs	Guidelines	Websites
Northwestern University	Use of Generative AI in Courses	https://ai.northwestern.edu/ education/use-of-generative-artificial-intelligence-in-courses.html
University of Texas at Austin	5 Things to Know about ChatGPT	https://ctl.utexas.edu/ teaching-technology/5-things-know-about-chatgpt
University of Wisconsin-Madison	Considerations for Using AI in the Classroom	https://idc.ls.wisc.edu/ls-design-for-learning-series/ considerations-ai-classroom/
University of York	Student Guidance on Using AI and Translation Tools	https://www.york.ac.uk/ students/studying/assessment-and-examination/ai/
University College London	Engaging with AI in Your Education and Assessment	https://www.ucl.ac.uk/ students/exams-and-assessments/assessment-success-guide/engaging-ai-your-education-and-assessment
University College London	Using Generative AI (GenAI) in Learning and Teaching	https://www.ucl.ac.uk/ teaching-learning/ publications/2023/sep/using-generative-ai-genai-learning-and-teaching

Continued

HEIs	Guidelines	Websites
London School of Economics and Political Science	LSE Short-Term Guidance for Teachers on Artificial Intelligence, Assessment and Academic Integrity in Preparation for the 2022–23 Assessment Period	https://info.lse.ac.uk/staff/divisions/Eden-Centre/Assets-EC/Documents/AI-web-expansion-Feb-23/Updated-Guidance-for-staff-on-AI-A-AI-March-15-2023.Final.pdf
Nanyang Technological University, Singapore	NTU Position on the Use of Generative Artificial Intelligence in Research	https://www.ntu.edu.sg/research/resources/use-of-gai-in-research
University of Tokyo	Policy on the Use of AI Tools in Classes	https://utelecon.adm.u-tokyo.ac.jp/en/docs/ai-tools-in-classes
Nagoya University	Regarding the Use of Generative AI	https://en.nagoya-u.ac.jp/academics/ai/index.html
University of Hong Kong	GenAI	https://tl.hku.hk/2024/08/35063/
Hong Kong University of Science and Technology	AI Literacy for End-Users: Use AI Wisely	https://libguides.hkust.edu.hk/ai-literacy
Chinese University of Hong Kong	Use of AI Tools in Teaching, Learning and Assessments: A Guide for Students	https://www.aqs.cuhk.edu.hk/documents/A-guide-for-students_use-of-AI-tools.pdf
City University of Hong Kong	Guidelines on Generative AI	https://www.cityu.edu.hk/GenAI/guidelines.htm

Continued

HEIs	Guidelines	Websites
Education University of Hong Kong	AI Generative Tools: Guidelines	https://libguides.eduhk.hk/ai-generative-tools/guidelines
Taiwan University	Guidance for Use of Generative AI Tools for Teaching and Learning	https://www.dlc.ntu.edu.tw/en/ai-tools-en/
Russell University Group	Russel Group Principles on the Use of Generative AI Tools in Education	https://russellgroup.ac.uk/news/new-principles-on-use-of-ai-in-education/
University of Edinburgh	Guidance for Students and Staff on AI Use	https://www.ed.ac.uk/ai/guidance
University of Edinburgh	Generative AI Guidance for Students	https://www.ed.ac.uk/bayes/ai-guidance-for-staff-and-students/ai-guidance-for-students
University of Manchester	Artificial Intelligence (AI) Teaching Guidance	https://documents.manchester.ac.uk/display.aspx?DocID=70286
Institut d'Études Politiques de Paris	ChatGPT: Sciences Po fixe des règles et lance une réflexion sur l'IA dans l'enseignement supérieur	https://www.sciencespo.fr/fr/actualites/sciences-po-fixe-des-regles-claires-sur-lutilisation-de-chat-gpt-par-les-etudiants/
University of Melbourne	Assessment and AI	https://melbourne-cshe.unimelb.edu.au/ai-aai/home/ai-assessment

Continued

HEIs	Guidelines	Websites
University of Sydney	AI in Education	https://canvas.sydney.edu.au/courses/51655
University of Queensland	Using AI in Your Assignments	https://guides.library.uq.edu.au/referencing/ai-tools-assignments
McGill University	Using Generative AI in Teaching and Learning	https://deptkb.mcgill.ca/display/TLK/Using+Generative+AI+in+Teaching+and+Learning
Katholieke Universiteit Leuven	Responsible Use of Generative Artificial Intelligence	https://www.kuleuven.be/english/genai
University of Auckland	The Use of Generative AI Tools in Coursework	https://teachwell.auckland.ac.nz/resources/generative-ai/ai-tools-in-coursework/
University of Liverpool	Generative Artificial Intelligence in Teaching, Learning and Assessment	https://www.liverpool.ac.uk/centre-for-innovation-in-education/digital-education/generative-artificial-intelligence/
University of Technology Sydney	Developing Guidelines for Generative AI in Your Subject	https://educationexpress.uts.edu.au/collections/artificial-intelligence-in-learning-and-teaching/resources/developing-guidelines-for-generative-ai-in-your-subject/

Continued

HEIs	Guidelines	Websites
Beijing Normal University & East China Normal University	Guide for Students on Using Generative Artificial Intelligence	https://www.ecnu.edu.cn/info/1094/67178.htm
Shanghai Jiao Tong University	Teacher's Guide for Regulating Student Use of Artificial Intelligence Tools	https://ctldnew.sjtu.edu.cn/news/detail/1143
ShanghaiTech University	Guide for Using Generative Artificial Intelligence	https://ai.shanghaitech.edu.cn/2024/0327/c14346a1093334/page.htm
School of Cultural Industry Management, Communication University of China	Initiative for Correct Use of Generative Artificial Intelligence by College Students	https://scim.cuc.edu.cn/_t482/2023/1025/c8345a213131/page.htm
School of Continuing Education, Communication University of China	AI Use Rules in Undergraduate Theses (Projects) for Continuing Education	http://mdedu.cuc.edu.cn/mdedu2021/2021content.aspx?id=15190
Hubei University	Notification on the Detection for 2024 Undergraduate Graduation Theses (Projects)	https://jwc.hubu.edu.cn/info/1061/7877.htm

Continued

HEIs	Guidelines	Websites
Shanxi Agricultural University	Notification on the Trial of AIGC Detection for 2024 Undergraduate Graduation Projects (Theses)	https://jwc.sxau.edu.cn/info/1006/8821.htm
Tianjin University of Science and Technology	Notification on the Trial of AIGC Detection for 2024 Undergraduate Graduation Projects (Theses)	https://jw.tust.edu.cn/ggtz/65b256ce743a41fa82c3533e16009807.html
Fuzhou University	Notification on AIGC Detection for 2024 Undergraduate Graduation Projects (Theses)	https://jwch.fzu.edu.cn/info/1039/13415.htm

Appendix 3　Usage Suggestions in "ChatGPT and Artificial Intelligence in Higher Education: Quick Start Guide"[①]

1. Use ChatGPT with Care and Creativity

i　Create opportunities for faculty, staff, students, and other stakeholders to discuss the impact of ChatGPT on the HEIs and co-construct strategies to adapt and adopt to AI.

ii　Introduce clear guidance for students and instructors about how and when ChatGPT can be used (and when it cannot).

iii　Connect the use of ChatGPT to course learning outcomes.

iv　Review all forms of assessment and evaluation to ensure that each element is fit for purpose.

v　Review and update policies relating to academic integrity/honesty in relation to ChatGPT and other AI tools.

vi　Train teachers, researchers, and students to improve the queries they pose to ChatGPT. As researchers have noted, ChatGPT is most useful when the inputs provided to it are carefully created.

2. Build Capacity to Understand and Manage ChatGPT

i　New programs/courses that focus on ChatGPT/AI will increase research and development capacity and provide students with cutting edge knowledge.

ii　Existing programs/courses can be updated to incorporate teaching

---

① UNESCO, "ChatGPT and Artificial Intelligence in Higher Education: Quick Start Guide," 2023, https://unesdoc.unesco.org/ark:/48223/pf0000385146.

of: AI literacy, AI ethics, and core AI competencies and skills.

iii   Training for staff can ensure that the support they provide to students and other stakeholders builds on rather than replicating what chatbots/AI tools offer and increase confidence in the deployment of technology.

iv   Peer support and mentoring for faculty members to increase skill level and share good practices for teaching and ways of using ChatGPT in research can be done within faculties, at institutional level, or among supra-institutional communities of knowledge.

3. Conduct an AI Audit

i   Understand the current situation. Understand what is data-driven AI, the types of data that need to be collected, and the data processing methods involved. Assess the types of AI currently used in HEIs and how they support institutional functions. Review existing policies or regulations at the institutional level, including those related to AI usage, privacy, and protection, as well as external regulations HEIs need to comply with.

ii   Decide which AI to use. Define the areas where AI could be applied, such as student services, assessment, or research, and determine the added value of AI technologies. Establish the HEI's position on open-source accessibility versus commercial AI tools and how to ensure inclusive access for all stakeholders.

iii   Monitor performance and equity. Evaluate the effectiveness of AI in meeting identified needs and set criteria for measuring its performance. Assess whether the collected data can be utilized by the HEI, including usage methods and data collection frequency. Determine the extent to which AI addresses equity concerns and how this is measured.